Climate Adaptation Policy and Evidence

Evidence-based policymaking is often promoted within liberal democracies as the best means for government to balance political values with technical considerations. Under the evidence-based mandate, both experts and non-experts often assume that policy problems are sufficiently tractable and that experts can provide impartial and usable advice to government so that problems like climate change adaptation can be effectively addressed; at least, where there is political will to do so. This book compares the politics and science informing climate adaptation policy in Australia and the UK to understand how realistic these expectations are in practice.

At a time when both academics and practitioners have repeatedly called for more and better science to anticipate climate change impacts and, thereby, to effectively adapt, this book explains why a dearth of useful expert evidence about future climate is not the most pressing problem. Even when it is sufficiently credible and relevant for decision-making, climate science is often ignored or politicised to ensure the evidence-based mandate is coherent with prevailing political, economic and epistemic ideals. There are other types of policy knowledge too that are, arguably, much more important. This comparative analysis reveals what the politics of climate change mean for both the development of useful evidence and for the practice of evidence-based policymaking.

Peter Tangney is a Lecturer in Science Policy and Communication at Flinders University, Australia.

Science in Society Series

Series Editor: Steve Rayner
Institute for Science, Innovation and Society, University of Oxford

Editorial Board: Jason Blackstock, Bjorn Ola Linner, Susan Owens, Timothy O'Riordan, Arthur Petersen, Nick Pidgeon, Dan Sarewitz, Andy Sterling, Chris Tyler, Andrew Webster, Steve Yearley

The Earthscan Science in Society Series aims to publish new high-quality research, teaching, practical and policy- related books on topics that address the complex and vitally important interface between science and society.

This excellent book provides an important exploration of the way that climate and adaptation science are done, and the way they can be, and are, politicised in each step of the process from science to the development of policy. Tangney's rich examination identifies the primacy of the legitimacy of science over its credibility and salience, and offers the significant conclusion that adaptation policy needs not only science, but also more explicit discussion of the normative priorities of government. Tangney offers policy makers an important rationale and roadmap for the development of adaptation policy.

David Schlosberg, Professor of Environmental Politics and co-Director of the Sydney Environment Institute, University of Sydney

This study of the role of scientific evidence in the practices of climate adaptation policy-making in the UK and Australia is shocking for those who still hold to a linear-technocratic model of science advice. Tangney shows in detail for this subject that more, supposedly better climate science does not necessarily speak well to the political problems encountered. With this book he adds a significant piece to the growing body of literature on the way in which climate adaptation problems are complex, uncertain and subject to diverging values and priorities.

Professor Arthur Petersen, Department of Science, Technology, Engineering and Public Policy (STEaPP), University College London (UCL)

This book provides a timely discussion of how science and politics interact in the evolution of adaptation policy frameworks. It is well known that the challenges presented by climate change are very broad, spanning many inter-related areas including disaster management, ecology, natural resources, infrastructure provision, human health and wellbeing. These activities all depend on both expert knowledge and lay perceptions about risks, responsibility and effective intervention. Tangney deftly characterises the modes of politicisation of policy knowledge and action that can occur in relation to climate change, and persuasively dismisses naïve forms of reliance on 'more science' to resolve adaptation issues.

Brian Head, Professor of Public Policy, School of Political Science, University of Queensland

Climate Adaptation Policy and Evidence

Understanding the Tensions between Politics and Expertise in Public Policy

Peter Tangney

LONDON AND NEW YORK

First published 2017
by Routledge

2 Park Square, Milton Park, Abingdon, Oxfordshire OX14 4RN
52 Vanderbilt Avenue, New York, NY 10017

Routledge is an imprint of the Taylor & Francis Group, an informa business

First issued in paperback 2018

British Library Cataloguing-in-Publication Data
A catalogue record for this book is available from the British Library

Library of Congress Cataloging-in-Publication Data
Names: Tangney, Peter.
Title: Climate adaptation policy and evidence : understanding the tensions
between politics and expertise in public policy / authored by Peter Tangney.
Description: Abingdon, Oxon ; New York, NY : Routledge, 2017. |
Series: Earthscan science in society series | Includes bibliographical
references.
Identifiers: LCCN 2017005829| ISBN 9781138284814 (hbk) | ISBN
9781315269252 (ebk)
Subjects: LCSH: Climatic changes--Political aspects. | Climatic
changes--Political aspects--Australia. | Climatic changes--Political
aspects--Great Britain. | Climatic changes--Social aspects. | Climatic
changes--Social aspects--Australia. | Climatic changes--Social
aspects--Great Britain.
Classification: LCC QC903 .T36 2017 | DDC 363.738/747--dc23
LC record available at https://lccn.loc.gov/2017005829

ISBN: 978-1-138-28481-4 (hbk)
ISBN: 978-0-367-15241-3 (pbk)

Typeset in Bembo
by Saxon Graphics Ltd, Derby

To A.L.

Contents

Figures and tables

Preface

In 2005, when I first moved to the UK, climate change was a hot topic amongst environmental scientists and policy analysts. Although adaptation was still often considered a 'cop-out' relative to the pressing need for greenhouse gas reduction, tentative steps were nonetheless being made toward the development of adaptation policies and plans. At that time, however, not much technical climate change impact and risk analysis was being done by anyone. Many believed that the arrival of the UK's next set of climate change projections would provide many of the necessary answers.

After a couple of years working as a consultant to UK water companies and the Department of Environment, Food & Rural Affairs, I moved to the Environment Agency of England & Wales, where I was enrolled as a national Climate Change Advisor. My task was to assess (and to help others to assess) climate change risks and, eventually, to interpret and disseminate the outputs of the UK's Climate Projections 2009 (UKCP09). My colleagues and I interpreted climate change largely in the abstract: some horrible spectre that no one had really seen, but that might wreak havoc on environmental management if we weren't prepared. It was during this time that I began to realise the full extent of the difficulties of using climate science for risk-based policymaking.

In 2011, I moved to Brisbane to undertake PhD studies, at a time when man-made climate change was temporarily in vogue with state and federal governments. I arrived in the aftermath of the worst flooding events in southeast Queensland in over 35 years; this time, 35 people were dead, along with AU$5 billion in damages. A sense of shock was particularly felt, perhaps, because the rains in advance of the floods had marked the end of a crippling nine-year drought in the region. I began to realise that contrasting chronic and acute climate extremes are not abstract phenomena for Queenslanders, as in the UK; they are often a pervasive reality.

In Brisbane I was fortunate to work on federally funded research projects commissioned by Australia's National Climate Change Adaptation Research Facility, and I spent a significant portion of my time canvassing the views of policy players at all levels of government involved in climate adaptation, disaster risk and emergency management. These encounters revolutionised my views about climate change risk and the task of adaptation policymaking. I began to realise that more and better science, as long fetishised in the UK, was not going to solve most of the adaptation policy problems I had encountered. These were political, *normative* problems, even when they were often couched in scientific or technical terms.

Climate change, apart from being an imminent threat to humankind, I believe, has also provided a unique opportunity to better understand societies' relationships with uncertainty and the natural environment; why these relationships can be so confounding for public decision-making, and the extent to which science and politics can effectively collaborate for that purpose. It may indeed make us stronger.

Peter Tangney
January 2017

Acknowledgements

This book would not exist were it not for the help of a few key people at Griffith University, to whom I will be forever grateful. In particular, I owe a debt to Michael Heazle and Michael Howes, who provided outstanding intellectual engagement during the development of my arguments and the analysis presented in this book. I also want to thank the various anonymous reviewers of earlier drafts for saying nice things about my work, giving me the confidence to submit a manuscript for publication. They hold no liability, however, for any gaffs or errors made here.

I thank Paul Burton and Jago Dodson who facilitated my arrival in Australia and provided me the opportunity and the intellectual freedom to pursue my research at the Urban Research Programme (URP) at Griffith University. Special thanks also to Michelle Lovelle and Anne Krupa at the URP for always being so supportive and helpful.

During the preparation of this book, I relied on a number of people in various ways, from reading draft chapters and providing formatting guidance, to help with preparing maps and figures; others graciously acted as a sounding board for my developing ideas, or provided similar hand-holding services. I give particular thanks to Tony Matthews, Angela Ballard, Michael Marriott, Darrin Durant and Josh Newman. Thanks also to everyone at the Centre for Science Education in the 21st Century at Flinders University, but especially to Martin Westwell, Diane Vomberg, Florence Gabriel, Ancret Szpak, Rachel Crees and Kristin Vonney.

Finally, I thank the great friends I have made in Australia, as well as my parents Noel and Patricia, for being generally lovely, especially for those times when I didn't deserve it.

Abbreviations

CCIRG	Climate Change Impacts Review Group
CDF	Cumulative Distribution Function
COAG	Council of Australian Governments
COP	Conference of the Parties
CPRS	Carbon Pollution Reduction Scheme
DEFRA	The UK Department of Environment, Food & Rural Affairs
DERM	Department of Environment and Resource Management
DRM	Disaster Risk Management
EA	Environment Agency of England & Wales
EC	European Commission
EEC	European Economic Community
EMA	Emergency Management Australia
ESD	Ecologically Sustainable Development
ESM	Earth System Model
ETS	Emissions Reduction Scheme
EU	European Union
GCM	Global Circulation Model / Global Climate Model
GDP	Gross Domestic Product
GHG	Greenhouse Gas
IPCC	Intergovernmental Panel on Climate Change
LNP	Liberal-National Party
NAP	National Adaptation Plan
NGRS	National Greenhouse Response Strategy
NPM	New Public Management
OFWAT	UK Water Services Regulatory Authority
OPEC	Organisation of the Petroleum Exporting Countries
PDF	Probability Density Function
PPRR	Prevent, Prepare, Respond, Recover
QRA	Queensland Reconstruction Authority
SEQ	Southeast Queensland
SEE	Southeast England
SES	Social-ecological System
STS	Science and Technology Studies

UKCCRA   UK Climate Change Risk Assessment
UKCIP   UK Climate Impacts Programme
UKCP09   UK Climate Projections 2009
UNFCCC   United Nations Framework Convention on Climate Change

1   Introduction

When reason is against a man, he will be against reason

Thomas Hobbes

Climate risk management is a persistent and challenging problem for policymakers around the world. Despite ongoing attempts by the scientific community to characterise climate-related problems in politically incisive ways, they continue to be exceedingly tricky to understand, define or to resolve in ways that are satisfactory to all (or even most) policymakers. Climate change, in particular, is a sustainable development issue that is so inter-related with almost every aspect of society, economy and ecology that policies need much more than simply scientific answers (Hulme, 2009). The rhetoric of both politicians and experts in recent years seems to reflect these difficulties. It has been described variously as 'the greatest and widest ranging market failure ever seen' (Stern, 2006), 'more serious even than the threat of terrorism' (King, 2004) and 'one of the greatest moral challenges of our age' (Rudd, 2009). Alternatively, it is 'absolute crap' (Abbott, 2009, cited in Rintoul, 2009), 'a Frankenstein monster that threatens to devour its own designers' (Peiser, 2009) and an 'expensive hoax' (Trump, 2014).

While the contentious nature of climate change has undoubtedly delayed efforts toward international agreement on greenhouse gas abatement, it has also posed serious difficulties for policymakers seeking to adapt to both existing and future climates. This book investigates these policymaking difficulties, and in particular, how experts and other policy players[1] understand climate, and thus, how *evidence-based* adaptation policies develop. Although often an ancillary issue for policymakers, climate adaptation has become an unexpected arena for debating fundamental norms and societal values and priorities, and not simply through the legitimate decision-making processes of democratic government. Contrary to prevailing ideals, I argue, important values can be inscribed and political positions can be decided during the development of supposedly impartial technical evidence.

In the following chapters I describe how scientific evidence has been developed and used to date for climate adaptation policymaking in Australia and the UK. Although I investigate the epistemology of climate *change* science in some detail, the purpose of this book is to investigate the means by which a broader range of climate-related scientific knowledge is delivered and used for public policy; a process which involves a range of both expert and non-expert policy players. Such is the nature of expert knowledge concerning climate and our adaptation to it, as well as policymakers' and other players' expectations for this evidence, it continues to be developed, interpreted, used and ignored in ways that can legitimise prevailing politics and that suppress explicit normative debate. These political influences, I suggest, are symptomatic of the unavoidable tensions that exist in liberal democratic government between expert authority – derived from claims of privileged impartial knowledge – and democratic political authority (Heazle and Kane, 2015). Here, I describe how these tensions find their most daring expression in complex and uncertain policy problems like those associated with climate adaptation and change.

For some time it has been suggested that effective knowledge systems for sustainable development (i.e. those processes that ensure the provision of adequate knowledge for decision-making) should aim to enhance evidence attributes such as credibility, salience, legitimacy and iterativity. Some have argued that these attributes should be pursued during the development of policy knowledge in a way that ensures an appropriate balance between the need for technically adequate objective expert authority (such that it exists) on the one hand, and evidence which meets the practical and normative requirements of decision-makers on the other. In this book I use these ideas about effective knowledge systems to investigate the tensions between expert and political authority. I explain when and how decision-making norms and prevailing political forces interact in the development, provision and use of evidence for climate adaptation.

Although credibility has traditionally been considered the priority attribute of effective technical evidence, I argue that, in the context of understanding and managing social-ecological problems like climate variability and change, evidence legitimacy is of principal importance. Legitimacy is dependent on the congruence of prevailing political values and priorities with those values inscribed in, or attributed to, that evidence. Here, I show how legitimacy has come to be the principal attribute that determines the effective interpretation, communication and use of expert knowledge for climate risk management. This finding reflects the political nature of many seemingly technical climate-related policy problems and the limited ability of experts to apprehend some wholly objective reality in relation to them. To date, however, legitimacy has not only been considered by experts as a secondary (or even redundant) attribute of effective knowledge, the concept has been defined and used in ambiguous and sometimes conflicting ways by those concerned with the use of science and expertise for policymaking. Here, I propose to reconcile concepts of legitimacy in a way that adequately accounts for the political acceptability needed to ensure that evidence is usable for policymakers, and thereby, to explain the importance of this attribute for effective policy evidence.[2]

This is not to suggest that evidence should not also be technically credible; of course it should. In the following chapters, however, I demonstrate how values are an inevitable component of expert evidence for policy, and the importance therefore of developing sufficient legitimacy for this evidence so that, in turn, it can be used to effectively inform and legitimise adaptation policies. In the absence of sufficient legitimacy, scientific research and expertise about climate change is susceptible to processes of *deliberate* politicisation that influence evidence outputs and their use in unexpected ways. However, allow me to distinguish at this point between two types of politicisation prevalent in the development of policy evidence for climate adaptation.

The first type, a *politicisation-by-process*, is both unintentional and may be, in some respects, an inevitable outcome of climate change evidence development, whereby, experts must make important value judgements and subjective decisions when developing climate and adaptation evidence and disseminating its outputs to decision-makers. For the science of *future* climate change and its presentation for policymaking, in particular, this normative input involves more than the 'epistemic values' choices associated with the agreed norms of good scientific practice (Douglas, 2004). During the development of climate change projections and assessments of climate vulnerabilities, impacts and risks, and in the choice of which evidence to use and how to interpret and present it, non-epistemic value judgements are required that can have important implications within the hyper-politicised sphere of climate-related policy.

This form of politicisation, although often necessary to ensure the coherence of scientific outputs, calls into question idealistic notions that continue to influence how experts and policymakers interact; for example, that there is such a thing as wholly impartial expertise (Douglas, 2009). Although it would be unfair to suggest that scientists are acting inappropriately when making such value judgements, and it seems likely that where they recognise such normative input scientists would reject any description of its contribution to policy evidence as 'politicisation', I argue that we must at least acknowledge the possibility of a politicisation-by-process. In the hyper-politicised realm of contemporary climate science development, and when aspiring to an ideal of disinterested scientific investigation, we must account for all of the normative constituents of climate-related science that could allow policy players to unintentionally inscribe their political positions upon supposedly impartial evidence.

It is not my wish here to undermine the scientific community or the science they do. Nor do I wish to conflate the underpinning bio-geophysical science of the climate change phenomenon and the empirical validity of the evidence for past and present climate change (about which there is considerable certainty), with the much more problematic scientific outputs describing *potential future* climate and its effects. What I seek is to better understand the appropriate roles for science and expertise in policymaking considering the normative constituents of this knowledge, and amidst considerable uncertainty over the trajectory of future climate change and its impacts. Moreover, in this book I am just as interested in discussing a second type of politicisation, a *politicisation-by-agency*. This type of politicisation occurs as a result of a direct intervention in evidence development when either expert or non-expert policy players seek to deliberately adjust the outputs or communication of research evidence in support of political priorities, and under the guise of supposed objectivity. To avoid confusion, unless otherwise stated, when referring to politicisation in the course of this book, I am speaking about such a politicisation-by-agency. In order to understand the tensions between expert and political authority within the hyper-politicised arena of climate change policymaking, this latter form of politicisation can be usefully considered in tandem with processes of *scientisation*, whereby, both experts and non-experts seek to suppress political debates during the policymaking process through their recourse to objective 'facts' and scientific expertise.

This book focuses its investigation of evidence-based climate adaptation on the comparable cases of the State of Queensland in Australia, and on England in the UK. These cases have been chosen because they have had similarly structured liberal democratic government and both have been world leaders in adaptation policymaking, albeit in very different ways. Whereas Queensland's adaptation is framed by its ongoing exposure to climate extremes, I argue, England's adaptation is framed by its vulnerability to future climate change. Through these contrasting frames, I invite readers to resist any automatic association of climate adaptation with anthropogenic climate change. Whereas humans may have only started to drastically influence the climate in the last few hundred years, we have always been adapting to climate. When considering the activities associated with adaptation policymaking, therefore, we must consider the context of existing climate variability and vulnerability, as well as any new threat posed by anthropogenic climate forcing. Moreover, climate adaptation involves more than just the building of infrastructure or policy frameworks to ensure the stability of a socio-economic status quo. As I show here, these adaptations can often be maladaptive. Effective adaptation must also foster 'social capital' associated with government and community adaptive capacity, to ensure the flexibility of society in response to future unforeseeable events.

Evidence doesn't always help

Examination of the case-studies used in this book reveals a curious irony relating to the pursuit of evidence-based policy in Australia and the UK, one that can tell us much about the tensions between expertise and politics. Despite prominent Australian politicians' scepticism about climate science, and despite Australia's tense relationship with the international community in relation to its perceived neglect of the climate change issue (Bourke, 2014b; White, 2014), climate adaptation-enhancing policy and practice in Queensland is comparatively well developed, while largely avoiding many of the difficulties that climate *change* policymaking has encountered in the UK.

Queensland's policies have developed primarily under the guise of concurrent priorities such as disaster risk, emergency management and urban planning. Under conservative administrations (for example, the Newman state government, 2012 – 2015), the absence of any mention of climate change from policy and practice was made possible, in part, by the extreme nature of Queensland's climates. State and local governments are already well-rehearsed in the management of extreme events and over many decades have been forced to learn much more about how vulnerable, exposed and resilient Queensland's society and economy is to climate extremes, compared to liberal democratic governments in more benign climate regimes. I argue, however, that the Australian approach to adaptation policymaking has also been less problematic because governments at all levels have been less enthusiastic in their embrace of evidence-based policymaking, despite their explicit adherence to this schema. Their reticence has allowed much greater freedom for political motives to be explicitly and openly pursued, while allowing greater flexibility in the extent to which instrumentally problematic sets of evidence are expected to inform policy decision-making. This political transparency has occurred alongside the ever-present need for adaptation policies to manage existing climate variability and extremes.

By contrast, the UK's strict adherence to the evidence-based mandate in recent years under the Climate Change Act (2008) and its advocacy for climate change projections and risk assessments as principal policymaking tools, has resulted in significant political decisions (for instance, in relation to the management of sea-level rise and flood risk) to be covertly addressed through the development of policy evidence. This politicisation has occurred, I argue, because their explicit debate in legitimate political forums would reveal important conflicts between the conclusions of experts and the prevailing political priorities of the executive. Thus, my comparative analysis of these cases points to some rather inconvenient truths about the usefulness and usability of climate change science and policy evidence when seeking to legitimately address climate adaptation problems.

Evidence-based policy as practised in the UK, I argue, is a double-edged sword. Although the use of evidence for policy is an important ideal, the type of legitimacy granted to evidence relating to contentious policy issues like climate change may actually result in a greater propensity for the politicisation of that knowledge under an evidence-based mandate. This is due to a fundamental mismatch between the 'rational' positivist[3] expectations of politicians and the public on the one hand, and the contingencies and intractable uncertainties of climate adaptation evidence and policy problems on the other. Under the evidence-based mandate, as currently pursued, the complex, contentious and uncertain nature of adaptation problems allows for some important political choices to be made in the most timely and politically expedient ways during the development of supposedly impartial expert evidence. To openly discuss these decisions in democratic forums might result in ongoing value debates and political conflict and inertia, whereas, under a *linear-technocratic* policy schema (see Chapter 2), I argue, such decisions can be usefully made under the guise of objective expertise through processes of politicisation.

In Queensland, by contrast, because the issue of climate change in recent years was deficient in both policy and evidence legitimacy, the adaptation policy process has proceeded primarily through governments' explicit political and ideological priorities alongside a selective adherence to the evidence-based mandate that could exclude, where necessary, contentious expert authority or any associated expectations for the role of climate change science in policymaking. I conclude that significant questions remain about the utility of problematic evidence sets such as climate change projections and expert risk assessment for adaptation policy, given the nature and extent of the uncertainties and associated political tensions that arise in relation to their use. These difficulties are demonstrated by the comparative cases in this research which show how, although climate change science is tactically and politically important for helping policy players to understand the broad nature of the problem and the potential scale of the associated hazards, this information is often simply too complex, uncertain and contentious to provide unequivocal direction or prescription. This evidence is fundamentally ill-suited to guiding policy action in a linear, instrumental way that could allow impartial or definitive assessment of future impacts and risk, or the optimisation of adaptation solutions. Moreover, this science is prone to deliberate politicisation when used in the development of policy evidence, in ways that are contrary to the ideals of liberal democratic policymaking. I conclude that evidence-based climate adaptation policy must pursue a different path, one that accepts the contingencies of both the expert evidence and the priorities of government.

Why is evidence-based climate adaptation policymaking problematic?

Scientists and economists often frame the problem of climate change in terms of CO_2 emissions, average changes in temperature or precipitation, or the loss of gross domestic product in order to communicate the immediacy of the associated risks and to elicit policy responses. However, these metrics are often less useful for informing policymaking beyond the development of its legitimacy as a worthy political priority. Commonly, this is because of a mismatch between, on the one hand, the global and regional scales at which this information is available, and the contextual and multi-scalar nature of climate adaptation policy issues on the other, for which much of the available science has limited value[4] (see Chapter 4). Where attempts at enhancing the resolution or precision of climate science have been made, this evidence is subject to increasing uncertainty and considerable difficulties of interpretation (Frigg et al., 2013b; Tang and Dessai, 2012; Dessai et al., 2009).

As I explain here, these difficulties are symptomatic of a broader issue concerning the extent to which technical experts can objectively apprehend and summarise the present and future state of society and the natural environment to provide useful assessments of potential environmental impacts or risk for policy. In particular, experts struggle to definitively characterise complex and uncertain climatic and 'social–ecological' systems[5] at a given level of governance, or to provide unequivocal judgement on context-specific considerations of climate risk and vulnerability. Resolving contemporary problems like climate change, therefore, presents a formidable challenge for government policymakers. Liberal democratic society has traditionally expected governments to look to experts, and in particular scientists and economists, to understand and inform such policy problems (Kitcher, 2011; Head, 2008a). However, the expert community has struggled with a problem notable as much for its political divisiveness as for its scientific and socio-economic intractability (Hulme, 2009).

Climate change cannot be understood or explained by scientists in the ways that politicians and the public have traditionally expected. Despite individual policy players' pragmatism in this regard, government prescriptions for evidence-based policymaking still seek to use evidence in a way that assumes that policy issues are sufficiently tractable and that the available evidence provides enough objective truth about adaptation problems to allow effective and rational solutions to be identified. These evidence-based approaches are based on a 'rationalist' linear-technocratic heuristic that assumes that policy is principally in need of scientific and technical evidence free from political influence in order to maximise social gain. As I argue here, adaptation problems defy this rational approach to policymaking because they resist definitive characterisation. Moreover, politics does not just occur in open democratic forums or during bureaucratic policymaking; norms and politics are also significant and inevitable determinants of evidence development. And, evidence-based adaptation policy is more prone to political influence precisely because it clings to rationalist ideals of objective instrumentalism that disguise the inevitable value-based components of evidence within the provision of objective facts. Thus, the linear-technocratic schema, I argue, facilitates the politicisation of adaptation evidence.

In this book I describe how both scientists and policymakers have, thus far, failed to adequately account for the influence of the norms, values and politics that are inextricably linked to the development of adaptation evidence. This failure, I argue, is indicative of positivist ideals that seek immutable scientific truths to guide action, but also, of ongoing attempts to achieve Enlightenment ideals of policy legitimacy that seek to make public decisions through a transparent and legitimate combination of expert and political authority. These ideals are based on the premise that there is a clear divide between these forms of authority and that they can work seamlessly together to arrive at rational policy decisions. I argue that such a clear interface between expert and political authority does not exist, for adaptation policy decision-making, or perhaps, at all. Experts must address normative/political questions in the development of evidence, just as non-expert policy players must provide important subjective and normative input to expert evidence to make it useful and usable for policy.

Thus, an ongoing tension exists within liberal democratic government: these forms of decision-making authority are expected to be separate and functionally differentiated, yet in order to function effectively they must also overlap.

The tendency for subjective and value-based interpretations to be inscribed within the apparent impartiality of expert evidence is certainly not a new problem for evidence-based policy approaches (Owens et al., 2004; Weale, 2001). However, this influence is, I argue, very significant in matters concerning the management of social-ecological systems, and in particular for climate adaptation due to:

• the complex, uncertain, multi-level and highly interconnected nature of individual adaptation problems;
• the pressing need to value future as well as present needs;
• the interdependence of climate adaptation policymaking with a broad range of concurrent policy priorities; and,
• the degree of political conflict climate adaptation issues provoke as a result of these characteristics.

The interpretive decisions required during the development of climate science and adaptation policy evidence highlight the difficulties of appropriately reconciling legitimate expert and political authority in the development of public policy. We may not ever be able to distinguish exactly where science ends and policymaking begins for such 'wicked' problems (Rittel and Webber, 1973) but, I argue, we cannot assume that this divide is located where experts and policymakers say it is. Both parties may have much either to gain or to lose by shifting the locus of this divide. In light of this, I argue here that legitimate and effective understandings of what is, what will be and what should be done to adapt to climate require a restructuring of how we think about and use science for adaptation policy, if not also for public decision-making more generally.

Australia and the UK: Two perspectives on the development and use of evidence for adaptation

Queensland, Australia and England in the UK provide intriguing comparative cases for the study of climate adaptation policymaking. As discussed above, the extent to which scientific evidence has been relied upon to justify adaptation policy is quite different in these two countries while their differing approaches sit in stark contrast to their experiences of climate to date.

Queensland's climate is a story of persistent exposure to contrasting extremes and society's ongoing need to adapt. Yet climate *change* has been a strongly polarising issue because of its tense relationship to the state's socio-economic development. By contrast, in England a relatively benign climate has been offset by increasing socio-economic vulnerability, a persistent public concern for the protection of the natural environment and bipartisan political concern for resource independence which ensures public and political support for prioritising the largely abstract risk of future climate disaster. What these two cases demonstrate is that, despite such differences in perspective, the political difficulties of deriving agreed solutions to adapt to climate and in agreeing about the suitability of the available evidence mean that politics has a tendency to overwhelm evidence development and expert authority during adaptation policymaking. Behind the approaches to climate adaptation taken by both Queensland and England lie tendencies to circumvent the evidence-based mandate where possible and where not, to disguise political decisions within technical evidence and regulatory metrics.

In Queensland, climate change has been subject to strong political polarisation such that its legitimacy as a priority political concern has always been in doubt. At the time of writing, the conservative coalition (Liberal-National) parties at all levels of government (federal, state and local), have rejected the climate change policies established by Labor governments past and present, due to differences in their socio-economic values and priorities. The federal government under Tony Abbott (2013–2015) began dismantling Labor's Clean Energy Act (2011)[6] as soon as they came to office, and at state level the Newman government (2012–2015) systematically dismantled Labor's policies and plans to explicitly adapt to climate change (Dedekorkut-Howes and Howes, 2014). Furthermore, concurrent policy initiatives such as disaster risk management and urban planning schemes, although capable of incorporating climate change science, generally do not do so as well as they might (or at all) due to a lack of legitimacy for both the policy and the science of climate change. At federal level, at the time of writing, there appears to be very limited strategic consideration of climate change,[7] and at state level the funding arrangements for disaster risk management policy also ensure a lack of incentive for states and local authorities to incorporate such considerations. In the aftermath of climate extremes, for instance, state government is usually only provided funding by federal government to rebuild infrastructure on a like-for-like basis.

This is not to suggest, however, that relative to the UK at least, Queensland lacks adaptive capacity or that adaptation efforts are necessarily deficient at the current time. Assertions to the contrary appear stuck in a rather narrow conception of what climate adaptation is and should be. As I argue here, climate adaptation is all too often concerned with a 'predict-then-act' mode of policy development that seeks adaptation primarily through governments' 'top-down' provision of infrastructure and spatial planning. On the contrary, I argue, although quite different in focus, Queensland's existing policy provisions make a significant contribution to adaptation through the enhancement of community resilience and the adaptive capacity of local government. Following the old adage that 'exposure breeds resilience', in many respects this contribution appears equivalent to those policymaking efforts in the UK made under the guise of climate *change* adaptation.

Nonetheless, the political battle over climate change rages on in relation to its legitimacy as a policy issue in Queensland, in particular for the mitigation of greenhouse gases (GHG). Climate adaptation – which is largely the responsibility of local and state government – has been a bipartisan priority only as long as policy action isn't justified on the basis of climate *change*, including the use of climate science which has varying levels of perceived legitimacy and credibility amongst policymakers. In effect, therefore, although Australian governments espouse the virtues of evidence-based policy, some types of evidence are deemed more legitimate than others. It appears that climate change evidence holds little sway in affecting disaster risk management, water resources management, urban planning and other climate adaptation-enhancing policy. Yet these policy priorities persist out of necessity due to the extreme nature of Queensland's climates. In particular, provisions for the integration of emergency services with disaster risk management, and for the enhancement of community preparedness and resilience to climate appear quite advanced relative to similar liberal democratic governments around the world.

By contrast, in the UK there has been considerable bipartisan support for climate change as a focus for policymaking. This is demonstrated by the pioneering Climate Change Act (2008) which provides a legislative mandate to reduce GHGs and to adapt to existing and future climate impacts. Support for climate change policies has been derived through bipartisan consensus on national priorities relating to resource independence and the status of Britain as a player on the world stage, as well as through popular public interest in the preservation of the natural environment. The global financial crisis in 2008–2009 and the European public debt crisis in its aftermath, however, have caused adaptation to be downgraded as a policy priority, irrespective of the legislative mandate this consensus previously created. Although support for action on climate change remains high and there is a strong bureaucratic attachment to the evidence-based mandate, successive governments have nonetheless sought, where possible, to prioritise present economic concerns over future climate risks.

At the time of writing, successive Tory (Conservative) national governments have removed responsibilities from local government to report on climate risks and adaptation policies. Instead, they have sought to achieve the requirements of the Climate Change Act (2008) through the existing reporting and regulatory mechanisms of government agencies and through voluntary agreements with major infrastructure and public service providers. Where there is legislative freedom to do so (e.g., at local government level) adaptation is now only justified on the basis of a 'business case' relating to costs and benefits in the present and immediate future, which supersede evidence about future climate risks, thereby effectively de-legitimising climate change evidence for local government policymaking.

Where government activity falls under the remit of the requirements of the Climate Change Act (2008) however, I describe how politics has infiltrated the preparation of evidence designed to inform policy. In particular, the UK Climate Change Risk Assessment (UKCCRA) (a five-yearly requirement of the Act) has been significantly influenced by political forces in its first iteration, whereby political positioning and prioritisation were disguised within an apparently independent and impartial technical assessment of climate risks to the UK. Such politicisation was possible, I argue, due to the fundamental mismatch that exists between the nature of climate change science and expertise on the one hand, and the economic–rationalist, linear-technocratic expectations of the evidence-based mandate on the other. Therefore, as I discuss in Chapter 3, although policymakers in the UK talk a good game in relation to climate change and the instrumental value of climate science, the provisions in place to adapt to present and future climates are not considerably more advanced, and in some ways appear less effective, than those in place in the comparative case of Queensland.

Why study the expert-political tensions in climate adaptation policymaking?

Adaptation is an interesting policy problem, epistemologically speaking, because of its broad political and policy scope, because of its 'wicked' characteristics that make it difficult to adequately understand and resolve (Rittel and Webber, 1973), and because of the types of expert knowledge available for informing it:

> The thing about adaptation is that you can hang it on anything, you know, anything you're interested in, anything that you care about, or any decisions you're going to have to make where you'll have to live with the consequences for years or decades or more, it applies.
>
> (UK Policy Player 1)

Climate adaptation is an issue that relates to a broad range of political goals and policy objectives at multiple levels of government and across multiple scales[8] of governance. It is relevant to many of the prevailing and contested politics and norms of society and economy, as well as to the resilience[9] of individual communities and government jurisdictions. Although adaptation is often considered in the context of local communities (Huitema et al., 2016) its effective management often requires cross-level governance. It requires both strategic and context-specific management and in this sense, I argue, climate adaptation may be described (albeit, somewhat crudely) as a 'glocal'[10] problem (Rhodes, 2006). The contrasting strategic and contextual characteristics of many climate impacts and climate change risks mean that characterising adaptation issues can be problematic for policymaking across multiple government levels. These characteristics also suggest, however, that the politics of the broader problem of climate change inevitably influence how governments understand and address climate adaptation.

Climate adaptation is relevant to almost everything humans do, and so climate adaptation policy relates to a large proportion of existing governance portfolios. Yet, unlike the broader issue of climate change, climate adaptation can be framed in ways that align more easily with existing political priorities and goals. As the case-studies in this book show, adaptation policymaking can claim allegiance to the evidence-based mandate, while still taking a partisan or sceptical position in relation to climate change science or its much-lauded expert consensus (Cook et al., 2013). Climate adaptation is a useful focus of study, therefore, because it is a problem that exemplifies the difficulties of negotiating the uncertainties and epistemologies of contemporary policy problems and that presents complex challenges for understanding the interactions and tensions between expertise and politics in the absence of definitive evidence about the future.

The ways in which Australia and the UK have constructed and seek to understand the climate adaptation problem tells us much about the influence of these countries' differing politics and socio-economic norms, goals and values on this policy issue. These priorities originate from contrasting historical and cultural development between these countries, which has resulted in varying attitudes and approaches to building climate resilience and ensuring present and future economic prosperity. These priorities relate to the political economy of natural resource use, public and political expectations for the permanence of society and its infrastructure, and the relative priority given to addressing and maintaining the resilience of socio-economic communities. The political divisiveness of anthropogenic climate change described by this comparative analysis sits in contrast to both cases' explicit adherence to the evidence-based mandate. Yet, these cases show the different ways the adaptation issue can be addressed using legitimised evidence, depending on governments' political priorities.

Climate adaptation, however, is also challenging for evidence-based policymaking due to the nature of the available evidence. Climate science often lacks salience for decision-makers because of the difficulties of using it to understand adaptation problems at both local and strategic scales, or for incorporating it into the most established tools of evidence-based policymaking. As I describe in subsequent chapters, these difficulties relate to understanding or addressing climate adaptation problems in a way that is meaningful and/or agreeable to all (or even most) relevant stakeholders.

It is due to these characteristics, I argue, that adaptation policymaking under an evidence-based mandate has become highly susceptible to becoming deliberately politicised. Effective decision-making seems to require an understanding of both the local and the strategic elements of the problem across multiple levels of government (local, state and national) and across multiple scales (geographic, temporal and jurisdictional). The technical, financial and logistical difficulties of achieving this mean that, instead, policymakers often present evidence from a limited range of perspectives, and usually only from a single governance level. As Cash et al. (2006: p. 6) argue:

> The drive to frame issues at a single level comes from the need to both simplify and control. Governments, for example, frame problems so that they become tractable within their jurisdictions.

As such, synoptic policy evidence necessarily incorporates a value position that manifests itself in the choice of evidence and the framing and characterisation of climate risks from a particular governance level or scale and according to prevailing politics; that is, policy evidence is constructed for political purposes. Under a linear-technocratic policy model which construes policy evidence as largely impartial and objective, I argue, this framing and characterisation can result in either the prioritisation of certain types of evidence over others in line with prevailing politics – as in Australia – or can result in the politicisation of a single consensus evidence base – as in the UK. These two modes of scientisation of the adaptation policy problem (see Chapter 7) present a significant challenge to government's development of 'robust' policy evidence as envisaged by the linear-technocratic schema, and therefore provide a useful focus for investigating the relative roles of expertise and politics in policymaking.

Understanding the politics of evidence-based policy

The fundamental mismatch between understanding the climate adaptation problem on the one hand, and the contrasting forces of rationalism and politics in bureaucratic government on the other, poses significant questions about how experts and policymakers can and should interact. The principal question this book seeks to answer, therefore, is: *How do political attitudes toward climate science influence the development of evidence-based climate adaptation policy?*

To answer this question, I examine the following prevailing assumptions in the development of climate adaptation policy:

- Adaptation policy is and should be entirely evidence-based, and that evidence for adaptation should be expert-driven and principally scientific, technical and/or economic in content.
- The most important scientific evidence required for adaptation policy is robust and usable descriptions of the meteorological hazards and potential impacts from future climate change.
- There is heuristic value for policymaking in the notion that science informs policy in a rational and linear way.

As discussed above, these assumptions have developed from ideals concerning the importance of evidence and reason to inform policy, rather than relying solely on prevailing norms or political ideology. They also stem from an associated ideal epistemology that assumes that science can find discrete, impartial and linearly informative answers to how the world works and that this information can help us to make robust decisions about the future. However, as I demonstrate here, the climate change problem is characterised as much by its uncertainties and complexity as by the dramatic conclusions scientists make about the future and so adaptation policymaking has encountered significant difficulties when pursued on the basis of these linear-technocratic assumptions.

Adaptation policymaking cannot rely on the provision of deterministic (or even conventional probabilistic) and impartial expert evidence about future climate change; we should not assume that evidence development and use is free from political influence, nor that policy players are (or can be) impartial servants of bureaucratic government that can rationally oversee evidence-based policy. On the contrary, I argue that:

- In the face of intractable uncertainties, *climate change projections (i.e., climate science) and the assessment of potential impacts (i.e., policy evidence) often lack one or more attributes* of *salience, legitimacy* or *credibility* for policymakers and the public in helping them to adapt. Neither climate science nor derived policy evidence is salient because these outputs cannot meet the instrumental-rationalist norms of decision-makers expecting deterministic and objectively correct answers to policy questions. Both of these forms of evidence often lack legitimacy too when they conflict with prevailing political and socio-economic norms, values and priorities. Policy evidence in particular necessarily relies upon a narrow framing of what any given impact or risk from climate change may be, and therefore often fails to account for the multiple valid perspectives of climate risk in a way that is considered to be fair and representative. Further, synoptic policy evidence derived through *ex ante* policy appraisal also lacks credibility because it fails to account for the perspectives and priorities of policy players at other governance levels and scales.

- Although calls for more and better science have often been used as a means of demonstrating action on climate change in the absence of political legitimacy for more substantive policy measures, *more and better information about the dynamics and trajectory of future climate change is not the most important type of evidence needed to inform adaptation policy once the legitimacy of climate change as a policy issue has been established.*

- *Science and scientists do not and cannot inform adaptation policy in a rational or linear-technocratic way* because of the nature of bureaucratic policymaking processes, the wicked characteristics of the climate change problem, and the necessity for co-produced policy knowledge for effective adaptation to climate change. Adherence to this rationalist heuristic, I argue, can actually help to disguise political debates within 'objective' assessments of science for policy and thus, I argue, enhances the propensity for the deliberate politicisation of adaptation evidence. *This rationalist ideal facilitates the scientisation of adaptation policy making, whereby, recourse to the 'facts' suppresses important normative, political debates.*

- *Adaptation policy can only be evidence-based to the extent that the available evidence is congruent with political priorities.* The political legitimacy of climate change evidence is largely determined by the existence of political legitimacy within liberal democratic government for climate change as a priority policy concern. Legitimacy (interpreted as a political acceptability for both the knowledge production process *and* the resulting knowledge) therefore is the principal limiting factor for evidence use in adaptation policymaking.

As a result of these difficulties associated with prevailing assumptions underlying adaptation policymaking, I conclude that it is time to reconsider not only what evidence scientists should strive to deliver for adaptation policymaking, but how we value and interpret the use of expertise for informing policy more generally.

Organisation of the book

The chapters that follow comprise a combination of theoretical analysis and critique of science-policy interactions; a comparative political analysis of Queensland, Australia and England in the UK in relation to climate change policy; an epistemological analysis of climate change evidence and the 'wicked' characteristics of climate adaptation policy problems; an analysis of policy players' perceptions of evidence credibility, legitimacy and salience, and a series of comparative case-study analyses from southeast Queensland and southeast England that demonstrate the nature of adaptation policy problems and the political processes involved when governments seek to address them. Although this book can be read from beginning to end, some chapters are also relatively self-contained. For instance, Chapters 2, 3, 4 and 8 can be read on their own where a reader wants to know about, respectively, science-policy interactions; the politics and history of climate change in Queensland and the UK; the epistemology of climate science and adaptation problems, and the nature of climate resilience. Chapters 5 to 7, however, will require many readers to have absorbed previous chapters. As much as practicable, however, each chapter provides signposts to related sections in the book where readers can find further explanation.

Chapter 2 provides a critique of linear-technocratic perspectives of policymaking and the developing theoretical perspectives about the roles of science and expertise for informing policy. This chapter will provide readers with a foundational understanding of the concepts underlying evidence-based policymaking in liberal democracies and the tensions that exist between expert and political authority when attempting to manage contemporary public policy issues in an evidence-based way. I provide an overview of developments in the social studies of science to help the reader understand social-constructivist perspectives of what science is and what it can do. Finally, I discuss the contrasting roles for experts and non-expert policy players in adaptation policymaking.

Through comparative political analysis, Chapter 3 introduces the socio-political and historical background and context for Queensland and England in order to understand the prevailing attitudes of both politicians and the public to climate change as a policy priority. I also describe their relative approaches to adaptation policymaking since 2005 to understand how dynamic factors relating to partisan politics can significantly impact both the development and use of evidence and policy for this purpose.

Chapter 4 describes how evidence for climate change policy develops. I outline a multi-step 'co-production' process in which evidence progresses from the development of climate change models; to a negotiation about how those model outputs will be presented to those who might use them; to the development of impact and risk assessments whereby policymakers and experts negotiate a body of mutually agreed and internally consistent policy knowledge. This characterisation will help to explain how subjective and normative assumptions and decisions are a necessary component of climate and adaptation science due to the existence of both tractable and intractable uncertainties about bio-geophysical and social-ecological systems. Building on the theoretical critique in Chapter 2, I then assess the 'wicked' characteristics of adaptation policy problems that make their pursuit under an evidence-based mandate particularly prone to political influence.

Chapter 5 provides an in-depth discussion of the tensions that arise when developing effective science and other evidence for policy. For this purpose I focus on the knowledge systems framework developed by Cash et al. (2003). Since publication, this framework has been influential as a conceptual device by environmental science and geography scholars seeking to understand what makes effective evidence for policy. I discuss here the suitability of Cash et al.'s criteria of credibility, legitimacy and salience for describing the effectiveness of knowledge and knowledge systems for climate adaptation policy. Although I am a proponent of the framework as a practitioners' aide to evidence development, I propose an alternative interpretation of these criteria to enhance their utility for evidence-based policymaking.

In Chapter 6, I demonstrate the suitability of my revised form of Cash et al.'s knowledge systems framework using testimony from policy players in the UK and Australia involved in climate adaptation. My analysis examines the perceived efficacy of both climate change science and derived impact and risk assessments. Although this latter form of evidence is often not considered wholly scientific by the science community, it is nonetheless often used by the policymaking community (and interpreted by the media and the public) as impartial scientific evidence. Importantly, therefore, this chapter highlights the degree of overlap between the realms of expertise and political authority that occurs during policymaking.

Chapter 7 examines in detail the contemporary use of the rational linear-technocratic policy model and the *ex-ante* policy analysis tools used for climate adaptation in southeast Queensland and southeast England. I describe how this model can disguise political debates within science-policy analyses and I demonstrate how risk assessment outputs (in particular, the UK's 2012 Climate Change Risk Assessment and the 'Q100' flood risk metric in Queensland) have become politicised in the course of developing evidence-based policy due to the expectations of politicians and the public arising from the linear-technocratic schema. In the UK, synoptic risk assessment has been used as a means of legitimising political decisions regarding the relative priority of climate risks under the guise of impartial expert evidence, and yet was subsequently sidelined in favour of more overt political prioritisation. In Queensland, rationalist heuristics have also been used to disguise political deliberation and decision-making as expert evidence and has resulted in a technocratic approach to the concurrent policy priorities of disaster risk management and urban planning policy, resulting in a failure by some policymakers to account for the full range of risks presented by climate extremes.

Using these case-study regions, Chapter 8 discusses the potential evidence needs of adaptation policy players in practice. I provide a simple typology by which to understand the constraints that policymakers face when seeking to address adaptation policy problems and I use a series of examples that explain how uncertainty, complexity and political and contextual factors can be used to direct the development of evidence for effective policymaking. I then discuss the concept of climate resilience as a framework for understanding climate adaptation problems. I argue that a resilience-based framework shows considerable potential as a means to address the political obstacles to climate adaptation policymaking, compared to the risk-based decision frameworks commonly used or advocated to date.

Chapter 9 summarises the main arguments of the preceding chapters by posing and answering a number of questions relating to the problems of expert legitimacy and the influence of contextual and political factors on evidence development for policy. I end with some final reflections in which I suggest that although we should aim for a transparency of values in the development of evidence for adaptation policymaking, the linear-technocratic model has persisted precisely because it allows values and political priorities to pass undetected through the evidence development process, thus suppressing political debates and garnering tacit support for government positions in order to ensure the timely and expedient delivery of policy. While the linear-technocratic heuristic has pragmatic appeal in this regard, I argue that a more transparent relationship between politics and expertise can only be achieved by eschewing prevailing positivist ideals about the nature of science and expertise for social-environmental problems like climate change.

A note on theoretical frameworks

This book is written for climate adaptation scholars and practitioners and therefore is designed to be accessible to the widest range of expertise, including (and perhaps especially) for natural scientists. In doing so, I nonetheless pursue a constructivist epistemology that aligns with the emergent characteristics of climate science and the nature of adaptation policy problems, as described in some detail in later chapters. Although I may risk upsetting those of a positivist persuasion, adaptation policy problems, I argue, are too complex, uncertain and subject to diverging values and priorities to ever allow experts and scientific facts to account for their true nature in a reductionist or wholly instrumental way.

Critics of social constructivism have sometimes complained that those who pursue such an approach are resentful, anti-establishment relativists who deny the empirical validity of scientific facts and are eager to 'debunk for the sake of debunking' (Gross and Levitt, 1998: p. 26). As I seek to show here, such criticism is more than a little unjustified. Science can be both empirically valid and socially contingent in its derivation, interpretation and use. Recognising how and why does not delegitimise science for public decision-making; it makes it more robust and ensures its appropriate use in the public realm. Within the social sciences, however, constructivist frameworks can provide alternative interpretations of phenomena such as climate hazards and risks and the factual information describing them, which could give varying levels of recognition to, or acceptance of, their objective reality. These varying ideas about constructed knowledge broadly align with concepts of *weak* versus *strong constructivism*, as well as with ideas of *realism* and *critical realism*.

The analysis in this book is presented primarily from a *weak constructivist* perspective. Weak constructivism is based on the idea that hazards, such as those examined for this research, are objectively real; they have an ontological basis, and it is in attempts to apprehend this objective reality that experts and policymakers derive their version of this reality and assess risks. By contrast, the *strong constructivist* view argues that nothing is a hazard or risk in itself, but that anything can be, depending on one's perspective. Under this latter view, risks have no ontological basis or objective reality and we assign the label of risk to whatever we are concerned about (Lupton, 1999: p. 28). Much of the criticism levelled at social-constructivist critique is aimed at this strong form, on the basis that 'science works', and in most respects these criticisms, I believe, are broadly valid. Weak constructivist critique, meanwhile, has sometimes been trivialised as irrelevant or immaterial because it doesn't make any difference to how science is, or should be, used. As I seek to show here, however, the development and use of climate change science for policymaking suggests otherwise. The character of its normative content and the politics that surround this science means that it would be unwise to ignore or to trivialise weak constructivist arguments about the epistemology of climate science and expertise as it is used for public policy.

This book may also serve, in part, as a critique of *realist* perspectives that have long prevailed within the scientific community and which pursue an instrumental positivism which assumes that it is possible to objectively and wholly apprehend fundamental and immutable truths about the world using science. By contrast, the arguments contained herein align with a *critical realist* framework of analysis. I presume that both intransitive (that which knowledge is about, i.e., some objective reality) and transitive (constructed knowledge) components exist in the epistemology of climate adaptation evidence. As Roy Bhaskar (in Archer et al., 1998: p. xii) notes:

> The Western philosophical tradition has mistakenly and anthropocentrically reduced the question of what is to the question of what we can know. This

is the 'epistemic fallacy', epitomized by concepts like the 'empirical world'. Science is a social product, but the mechanisms it identifies operate prior to and independently of their discovery (existential intransitivity). Transitive and intransitive dimensions must be distinguished. Failure to do so results in the reification of the fallible social products of science. Of course being contains, but it is irreducible to, knowledge, experience or any other human attribute or product. The domain of the real is distinct from and greater than the domain of the empirical.

A critical realist approach is necessary, I argue, given the nature of contemporary climate change and adaptation science which uses complex models as a virtual laboratory through which to understand the climate system (Kellow, 2009), alongside co-produced evidence products that describe extant climate threats (Jasanoff and Wynne, 1998). These climatic and social-ecological models necessarily require significant subjective and normative interpretations to derive coherent scientific outputs. In this book, therefore, my weak social constructivism broadly aligns with this *critical realist* perspective. This perspective assumes that, even though expert and political forms of decision-making authority are distinguishable in terms of their methods, goals and objectives, there is no clear line of demarcation at the interface between them. Expert authority and political authority are in a constant state of tension and overlap. Although experts pursue *what was*, *what is* and *what can be*, and politicians pursue *what should be*, such is the indeterminacy and uncertainties concerning socio-cultural knowledge about the world, it can be difficult to discern where one pursuit ends and the other begins.

Notes

1 Throughout this book I use the term 'policy player' to correspond to the full range of expert and non-expert actors involved in the development and implementation of policy across all levels and institutions of government. By contrast the term 'policymaker' is only used in reference to those democratically elected to government.
2 Throughout this book I use the terms *climate science* and *policy evidence* in precise ways to mean two distinct forms of policy knowledge that reflect the character of evidence development for adaptation policymaking (see Chapters 2 and 4 for details). Where I use the word *evidence* on its own I am referring to the broader body of information available to policymakers. Likewise, the terms *political legitimacy*, *evidence legitimacy*, *credibility* and *salience* are used in precise ways (see Chapter 5).
3 'A philosophical system recognizing only that which can be scientifically verified or which is capable of logical or mathematical proof ...' (*OED*, 2017).
4 Wilbanks and Kates (1999) were perhaps the first to highlight these difficulties.
5 'A social-ecological system consists of a bio-geophysical unit and its associated social actors and institutions. Social-ecological systems are complex and adaptive and delimited by spatial or functional boundaries surrounding particular ecosystems and their problem context' (Glaser et al., 2008).
6 Containing a hard-won and controversial Carbon Emissions Trading Scheme (see Chapter 3).

7   Note, for instance, the absence of climate change or renewable energy from the Turnbull government's 2015 Science and Innovation Agenda (Australian Government, 2015).
8   For the purposes of the arguments in this book, *scale* refers to the spatial, temporal or otherwise quantitative or analytical dimensions used to measure governance. *Level* is the unit of analysis located at different positions on any given governance scale. Therefore, for instance, 'local', 'national' and 'international' describe levels of governance on a jurisdictional and/or geographical scale. I will also refer here to a temporal scale of governance and its associated levels relating to political electoral cycles (i.e., 3 or 4 years), decadal climate influences (i.e., 10+ years), as well as longer term climate change (30+ years).
9   See Chapter 8 for a discussion of the varying concepts of resilience and their relevance to adaptation evidence use and decision-making.
10   '... the meeting, intersection, overlap and coexistence of the particular and the universal' (Rhodes, 2006: p. 81).

Bibliography

Archer, M, Bhaskar, R, Collier, A, Lawson, T and Norrie, A (eds) 1998, *Critical Realism: Essential Readings*, Routledge, London and New York.

Australian Government, 2015, *National Science & Innovation Agenda*, Commonwealth of Australia, viewed 19 August 2016, www.innovation.gov.au/page/national-innovation-and-science-agenda-report/.

Bourke, L 2014b, 'G20 Summit: Barrack Obama puts climate change at fore in speech at University of Queensland', *The Sydney Morning Herald*, 15 November 2014, viewed 30 January 2015, www.smh.com.au/federal-politics/political-news/g20-summit-barack-obama-puts-climate-change-at-fore-in-speech-at-university-of-queensland-20141115-11ndmg.html/.

Cash, DW, Adger, WN, Berkes, F, Garden, P, Lebel, L, Olsson, P, Pritchard, L and Young, O 2006, 'Scale and cross-scale dynamics: Governance and information in a multi-level world', *Ecology and Society*, vol. 11, no. 2, viewed 16 March 2015, www.ecologyandsociety.org/vol10/iss2/art9/.

Cook, J, Nuccitelli, D, Green, SA, Richardson, M, Winkler, B, Painting, R, Way, R, Jacobs, P and Skuce, A 2013, 'Quantifying the consensus on anthropogenic global warming in the scientific literature', *Environ. Res. Lett.*, vol. 8, pp. 1–8.

Dedekorkut-Howes, A and Howes, M 2014, 'Climate Adaptation Policy and Planning in South East Queensland', in Burton, P (ed.), *Responding to Climate Change: Lessons from a Hotspot*, CSIRO Publishing, Melbourne.

Dessai, S, Hulme, M, Lempert, R and Pielke Jr, R 2009, 'Climate Prediction: A Limit to Adaptation?', in Adger, WN, Lorenzoni, I, and O' Brien, KL (eds), 2009, *Adapting to Climate Change: Thresholds, Values, Governance*, Cambridge University Press, Cambridge.

Douglas, H 2004, 'The irreducible complexity of objectivity', *Synthese*, vol. 138, no. 3, pp. 453–473.

Douglas, H 2008, 'The role of values in expert reasoning', *Public Affairs Quarterly*, vol. 22, no. 1, pp. 1–18.

Douglas, H 2009, *Science, Policy, and the Value-Free Ideal*, University of Pittsburgh Press, Pittsburgh, PA.

Frigg, R, Smith, LA and Stainforth, DA 2013b, 'The myopia of imperfect climate models: The case of UKCP09', *Philosophy of Science*, vol. 80, no. 5, pp. 886–897.

Gross, PR and Levitt, N 1998, *Higher Superstition: The Academic Left and its Quarrels with Science*, pbk edn, The Johns Hopkins University Press, London.

Head, B 2008a, 'Wicked problems in public policy', *Public Policy*, vol. 3, no. 2, pp. 101–118.

Heazle, M and Kane, J 2015, *Policy Legitimacy, Science and Political Authority: Knowledge and Action in Liberal Democracies*, Earthscan, London and New York.

Huitema, D, Adger, WN, Berkhout, F, Massey, E, Mazmanian, D, Munaretto, S, Plummer, R and Termeer, CCJAM 2016, 'The governance of adaptation: Choices, reasons, and effects. Introduction to the special feature', *Ecology and Society*, vol. 21, no. 3, pp. 37ff.

Hulme, M 2009, *Why We Disagree about Climate Change: Understanding Controversy, Inaction and Opportunity*, Cambridge University Press, Cambridge.

Jasanoff, S and Wynne, B 1998, 'Science and Decision-making', in Rayner, S, and Malone, EL (eds), *Human Choice and Climate Change*, Volume 1: *The Societal Framework*, Batelle Press, Columbus, OH.

Kellow, A 2009, *Science and Public Policy: The Virtuous Corruption of Virtual Environmental Science*, Edward Elgar Publishing, Cheltenham, UK.

King, D 2004, cited in The Carbon Brief, n.d., *Profiles – David King*, viewed 18 March 2015, www.carbonbrief.org/profiles/david-king/.

Kitcher, P 2011, *Science in a Democratic Society*, Prometheus Books, New York.

Lupton, D 1999, *Risk*, Routledge, London.

Owens, S, Rayner, T and Bina, O 2004, 'New agendas for appraisal: Reflections on theory, practice, and research', *Environment and Planning A*, vol. 36, no. 11, pp. 1943–1959.

Oxford English Dictionary [OED] 2017, 'Positivism', Oxford University Press, viewed 13 January 2017, https://en.oxforddictionaries.com/definition/positivism/.

Peiser, B 2009, *Quotes on Climate Change, Environment and Energy*, Climatism website, viewed 30 December 2014, www.climatism.net/quotes-on-climate-change-environment-and-energy/.

Rhodes, R 2006, 'Glocal: A few words on a theme', *Canadian Art*, vol. 23, no. 3, pp. 80–81.

Rintoul, S 2009, 'Town of Beaufort changed Tony Abbott's view on climate change', *The Australian*, 12 December, viewed 18 March 2015, www.theaustralian.com.au/archive/politics/the-town-that-turned-up-the-temperature/story-e6frgczf-1225809567009/.

Rittel, HW and Webber, MM 1973, 'Dilemmas in a general theory of planning', *Policy Sciences*, vol. 4, no. 2, pp. 155–169.

Rudd, K 2009, 'Earth Hour 2009', speech delivered 26 March, viewed online 31 March 2017, https://pmtranscripts.pmc.gov.au/release/transcript-16476./

Rudd, K 2013, '1st Leaders Debate for the 2013 Federal Election,' opening speech broadcast on ABC, 11 August, viewed 13 August 2013, www.abc.net.au/iview/#/series/12771/.

Stern, N 2006, *Stern Review on the Economics of Climate Change*, HM Treasury, London.

Tang, S and Dessai, S 2012, 'Usable science? The UK climate projections 2009 and decision support for adaptation planning', *Weather, Climate and Society*, vol. 4, no. 4, pp. 300–313.

Trump, DJ 2014, 'Snowing in Texas and Louisiana, record setting freezing temperatures throughout the country and beyond. Global warming is an expensive hoax!' Twitter feed: @realDonaldTrump, 28 January 2014, viewed 13 December 2016, https://twitter.com/search?f=tweets&q=from%3Arealdonaldtrump%20warming%20hoax&src=typd/.

Weale, A 2001, 'Deliberative democracy: Science advice, democratic responsiveness and public policy', *Science and Public Policy*, vol. 28, no. 6, pp. 413–421.

White, A 2014, 'Climate change "off the G20 agenda" as Australia prepares to abolish carbon price', *The Guardian*, 5 June 2014, viewed 30 January 2015, www.theguardian.

com/environment/southern-crossroads/2014/jun/05/g20-climate-change-agenda-obama-abbott/.

Wilbanks, TJ and Kates, RW 1999, 'Global change in local places: How scale matters', *Climatic Change*, vol. 43, no. 3, pp. 601–628.

2 Science, evidence and public policy

> The ideologies of our time (economism, scientism and technocracy) support the progressive view that experts, using scientific methods, can manage the world's problems by objective and efficient means ... Several aspects of that view are no longer tenable
>
> Donald Ludwig

This chapter provides a brief introduction to the complex and sometimes conflicting scholarly perspectives on the role of evidence and expertise in policymaking. This discussion will help readers unfamiliar with this literature to understand later chapters that describe the contributions of experts and evidence to climate adaptation, and how political attitudes toward their roles in policymaking can influence policy outcomes.

Evidence-based policy has taken on increasing significance for liberal democratic government at the beginning of the twenty-first century, though its origins date back to the Enlightenment or earlier. More than ever before, commentators would have us believe, we are producing vast quantities of evidence and we increasingly rely on expertise to justify, inform and otherwise influence the development of public policy[1] (Grundmann, 2009; Keller, 2009: p. 2; Nilsson et al., 2008; Maasen and Weingart, 2005; Jasanoff, 2003a). In the context of this prolific production of evidence for policy, however, there has also been a notable shift in policymaking practice concerning how this resource is developed and used. Increasingly, although often implicitly, scientists are expected by policymakers to act within a broader concept of expertise that draws upon their practical and analytical skills to cast judgement on issues that often transcend any individual's disciplinary competence (Fischer, 2009; Jasanoff, 2005). Moreover, traditionally certified experts must share their authority with a broader range of valid expert perspectives in the development of policy knowledge (Maasen and Weingart, 2005). This shift away from science as the sole arbiter for defining and understanding technical problems for policymaking has occurred despite the tenacity of technocratic concepts of scientific governance (Irwin, 2006), which still largely assume that science can 'speak truth to power' (Price, 1965). This latter concept is still reflected in the perspectives of many policy players (Grundmann, 2006) as well as in much of the rhetoric of, and guidance to, the policymaking community, advocating for 'sound science' as impartial and capable of apprehending objective reality (Hertin et al., 2009), sometimes even alongside calls for the democratisation of expertise (Irwin, 2006; Jasanoff, 2005).

Clearly, therefore, there are conflicting influences in the development of evidence-based policy which are useful to understand before investigating the particular case of climate adaptation policymaking. I begin here by discussing one of the most influential ideas in democratic policymaking since the nineteenth century: *linear technocracy* and its variants. I then describe some of the tensions between expertise and political authority that can arise due to the imposition of linear-technocratic ideals due to their failure to account for both the nature of policymaking and the expert knowledge those ideals are based upon. Understanding governments' pursuit of these ideals will also help to explain the difficulties associated with demarcating exactly where expert authority ends and politics begins during the policymaking process, to be described in later chapters. Finally, I describe some of the potential roles for evidence and experts relative to non-expert policy players, and two of the most fundamental problems that arise at the science-policy interface.

Jasanoff (2005) suggests that there are three bodies of expertise relevant to understanding the interactions between science and policy and in seeking answers to the aforementioned questions:

1   The bodies of knowledge that experts represent.
2   The experts themselves.
3   The committees and institutions through which experts offer judgement for policymaking.

This book focuses principally upon the first two of these bodies, in as much as it is meaningful or possible to assess their contributions without also discussing the ways in which they are organised. In the course of investigating the role of knowledge and expertise in policymaking, however, this chapter will show just how dependent effective evidence for policy is on the mechanisms and institutions of contemporary bureaucratic government.

Given the historical development of social and institutional legitimacy for evidence-based policymaking (Kitcher, 2011), it is unsurprising that much of the available theory described here is subject to multiple interpretations across sub-disciplines of political and social science (Spruijt et al., 2014; Miller and Neff, 2013). This chapter does not attempt a definitive summary; rather, I seek to consolidate varying theoretical interpretations and to find common ground across these literatures, in a way that is relevant for understanding the dynamics of adaptation policymaking and the role of experts and evidence therein.

Linear technocracy

Beyond the standard model which views policymaking purely in terms of the mechanisms of electoral transition (Jones and Baumgartner, 2012), there is a wide variety of models describing how policymaking ensues. Understanding some of these models can be useful for conceptualising how science and expertise are used for policy. For this discussion, however, I focus on an idea that continues to be a central assumption in the practice of evidence-based policymaking, even if scholarly theory has moved far beyond it.

The linear or 'rational' model has long held intuitive appeal and has been regarded as probably the most widely applied heuristic for policy decision-making (Althaus, Bridgman and Davis, 2007; APSC, 2007; Bell, 2004; Jasanoff and Wynne, 1998). Perhaps as a result of its persistent hegemony over policymaking theory in practice, the rational model has been the focus of extensive criticism and debate and it's probably fair to say that it has been thoroughly debunked as both a descriptive and a normative model by various academic literatures (Durant, 2016; Head, 2013; Keller, 2009; Grundman, 2009; Sutton, 1999). Nonetheless, both the interpretations and criticism of linear rationality can sometimes be challenging to untangle, coming as they do from all directions within the academy.

Although it is a broad theoretical frame, all variants of this model share a perspective that evidence can provide facts and apprehend objective reality with sufficient detail and accuracy to linearly inform how best to achieve political ends (Keller, 2009; Sabatier, 1999: p. 6). In political theory for instance, such interpretations include the 'rational actor' model where decisions are made by a unitary actor (Allison, 1971), and the 'rational-comprehensive' model where rational decisions are made through a linear, sequential set of steps each time a new policy response is required (Parsons, 1995: p. 271). Generally speaking, the broader rational decision-making model prescribes an approach which assumes that the decision-maker(s) is motivated to maximise social gain; that he or she has adequate objective knowledge of the problem and the value preferences of those involved, knows all the relevant options available, understands the costs and benefits of these options, and can rationally choose and evaluate the best one (Dye, 2005: p. 15). Sutton (1999) outlines the steps prescribed by the rational approach, as follows:

- recognising and defining the nature of the issue to be dealt with;
- identifying possible courses of action to deal with the issue;
- weighing up the advantages and disadvantages of each of these alternatives;
- choosing the option that offers the best solution;
- implementing the policy;
- possibly evaluating the outcome.

Across academic disciplines the broad characteristics of rationalism have also been subsumed under various related concepts such as 'Instrumentalism'

(Parsons, 2004) or 'Deficit Theory' (Irwin, 2006; Bucchi and Neresini, 2008). All refer to the expectation that the principal limiting factor for policy development and implementation is appropriate information or technical assessment which experts can and should provide in an objective and timely manner for rational decision-makers.

Amongst contemporary theorists, Keller (2009: p. 29) and Grundman (2006) highlight an important distinction between *rationalist (or linear)* and *positivist (or technocratic)* models; a distinction that is often lost in theoretical discussions of the use of evidence for decision-making as scholars refer to one or the other when actually talking about some combination of both, or vice versa. Rationalist policymaking, they argue, relies on normative or political choices to define the goals and priorities of policymakers, whereupon science then provides adequate technical answers as to how those ends may be achieved, or it identifies issues associated with their technical feasibility. By contrast, positivist policymaking is predicated on the view that science can resolve what might otherwise induce political debates by providing definitive answers about the correct allocation of resources and can therefore resolve conflict over policy choices and even goals. This latter model also assumes that the principal limiting factor of evidence-based policy is robust information, yet leaves even less room for the possibility of friction between the political and the scientific elements of policymaking. The positivist model advocates a form of technocracy justified on the basis that science can apprehend objective reality, not just in terms of how to achieve a given policy goal, but also in terms of defining that goal in the first place. This technocratic view of policymaking suggests that 'if scientific uncertainties are resolved, political debate will follow suit' (Keller, 2009: p. 29). In its contemporary form, Keller believes that the positivist model manifests itself as 'soft positivism' (Keller, 2009: p. 29), whereby policy players believe that science can sufficiently constrain the scope of political choice so as to make it significantly easier to reach policy consensus.

As Keller (2009) also notes, however, the positivist ideal appears as unviable as rationalism during incremental policy processes[2] that are principally motivated by the pursuit of consensus formation rather than on apprehending reality as a pre-requisite for policy decisions. Nonetheless, as this book seeks to demonstrate, a rationalist-positivist model, or perhaps more appropriately the *linear-technocratic model*, is particularly prevalent for understanding policy debates concerning climate change and climate adaptation policy, since it is still used as the default heuristic for evidence-based policymaking and still represents the views of many (though not necessarily most) policy players about the validity and potential utility of climate science and the reducibility of adaptation policy problems.

Although there is much that is attractive about this model as a heuristic device for practitioners (Heazle et al., 2013) it has also been highly problematic for policy decision-making. This is due to difficulties associated with agreeing on the goals that would dictate decision-making criteria; the nature of contemporary policy problems which often defy scientific or political agreement concerning what the problem actually is (or indeed, that there is even a problem in the first place); and the difficulties of objectively understanding what a 'best' response might look like (Head, 2008a,b; Nilsson et al., 2008; Sutton, 1999; Rittel and Webber, 1973). In later chapters, I discuss these difficulties in considerable detail.

Possibly the most notable proponent of rational decision-making was Nobel laureate Herbert Simon (cited in Parsons, 1995: p. 273) who nonetheless highlighted a significant limitation of the model. He suggests that decision-makers are ultimately subject to what he called 'Bounded Rationality'; the bounds of any given problem are far greater than the ability of that person or group to conceive of them. There are always so many alternatives, so much information to process that a wholly rational objective decision – one that is logically informed by all the relevant evidence – is not possible. Simon suggests that decision-makers are rational to the extent that they can conceive of a problem, within the bounds of cognitive ability, scientific understanding and the organisational environment in question. Bounded rationality is just one example of the many criticisms attempts at rational decision-making have provoked. More recently, an important criticism of linear-rationalism, and related to the issue of bounded rationality, has been the characterisation of many policy issues, including climate change (Head, 2008a,b), as *wicked*.

The concept of wickedness, first proposed by Rittel and Webber (1973), refers to the difficulties of definitively understanding very much at all about contemporary policy problems, as well as the irreducible, necessarily subjective and therefore political nature of any decisions made in relation to their management, or, in the use of evidence for that purpose. As I discuss in Chapter 4, wicked problems are highly interconnected, complex, uncertain and therefore difficult to understand, to characterise or to resolve in the ways traditionally expected in the development and delivery of scientific questions and answers. Thus it is argued, the aforementioned steps of rational policymaking cannot be completed sequentially in an impartial way that could ensure a transparently rational approach (Hertin et al., 2009; Head, 2008b; Oreskes, 2004; Sutton, 1999; APSC, 2007). In any case, evidence performs many more roles than the simple linear problem-solving one ascribed by the rational model. Evidence can also be used in tactical and political ways to advocate for policy actions without focusing on, or needing to understand, the precise technical details of that knowledge (Hertin et al., 2009; Weiss, 1979).

In a similar manner to distinctions between rationalism and positivism, and as I describe in more detail in Chapter 7, the linear-technocratic model is related to two further ideas concerning the interaction of expertise and politics – *decisionism* and *technocracy* – that are often inscribed within government guidelines and decision-making prescriptions for understanding and managing environmental risks such as those presented by climate variability and change. Irrespective of what interpretation of experts' role is assumed, however, the broader linear-technocratic model is problematic for the development of adaptation policy because it incorrectly assumes the availability of (if not value-free, then at least) disinterested, adequate and readily salient expert knowledge during the policy process. As I demonstrate in later chapters, these assumptions are highly questionable and can actually allow for significant political priorities to be determined covertly during the development of supposedly objective expert evidence and its supposed linear application to decision-making.

In light of these, and other criticisms, a variety of alternatives to linear-rationalism and its variants have been used to describe how policy is and should be made. These include, amongst others, the incremental model (Lindblom, 1959), the mixed-scanning model (Etzioni, 1967), the punctuated equilibrium model (Jones and Baumgartner, 2012), Actor-Network theory (Sismondo, 2008) and the bureaucratic politics model (Allison, 1971). All have their merits and all can be helpful for understanding the intricacies of policymaking. Perhaps the most important distinction to note at this point, however, is the extent to which these varying perspectives rely on assumptions about evidence-based rationality versus political deliberation and influence during decision processes.

For instance, Lindblom's (1959) description of incremental policymaking suggests that the policy process may be strongly driven by political interaction. Incrementalism – whereby policy is made through small incremental changes – provides a limit to the information requirements of policymakers while making it easier to reach consensus on a preferred policy option; the required steps are considerably less significant and therefore less contentious than might be required under a rational-comprehensive approach. Perhaps as prevalent for the arguments in this book, however, Lindblom (1959: p. 81) proposed that issue identification (along with the clarification of values, goals and priorities implicit in this) is actually a process that runs concurrent to and is elaborated by the analysis of options and the choice of a preferred policy, rather than a required step that precedes rational choice. In other words, the discrete steps set out in the linear-technocratic model do not occur in reality. Decisions about means often precede those about ends, and values and objectives are clarified along the way. As we will see in Chapters 7 and 8, this idea is often reflected in the practice of adaptation policymaking.

Policymaking as politics

Although Lindblom acknowledged political influence in policymaking, he also appeared to assume that impartial technical analysis and rational decision-making have a role to play. By comparison, Allison (1971) suggests that it may not be possible to understand the exact processes by which policy is made. He describes a Bureaucratic Politics Model that considers decision-making as a process of political bargaining amongst a range of decision-makers within government hierarchies:

the … Bureaucratic Politics Model sees no unitary actor but rather many actors as players … who make government decisions not by a single rational choice but by the pulling and hauling that is politics.

Allison (1971: p. 144)

Under this model, evidence appears to be a secondary or incidental concern. Policy decision-making is shared between a set of decentralised government actors, usually under the control of a central authority, who may disagree about what needs to be done, thus necessitating a political process whereby one group of actors committed to a particular course of action, triumphs over another group; or equally likely, different groups pulling in different directions produce a result distinct from what any individual or group intended. Similarly, the Game Model of policy delivery, advanced by Bardach (1977; cited by Parsons, 1995: p. 470) conceptualises government organisation as structures of groups and individuals seeking to maximise their power and influence. Actor-Network theory, meanwhile, is a broader social theory developed in the 1980s with similar implications for how evidence is developed and used. It suggests that the development and use of science for policy are the result of heterogeneous networks between both human and non-human entities (Sismondo, 2008). These various ideas have similarities to political and public administration theories associated with networked governance (see Howes et al., 2015).

Others have sought to explain policymaking as being more or less determined by historical and established socio-economic norms, values and structures (known as *structuralism*). Sabatier (1988) in particular is noteworthy because he builds upon the ideas of policy actors in networks to suggest that policymaking is influenced by a combination of stable and dynamic socio-economic and political structures and conditions alongside political interactions and relationships between policy players. Exogenous 'macro' social and economic factors combine with political machinations within policy '*subsystems*' relevant to specific fields of interest to decide policy means and ends. Policy subsystems comprise various '*advocacy coalitions*' which form and compete on the basis of shared values, beliefs and resources. The membership of these coalitions includes, not just administrative or government sponsored agencies and interest groups, but players from all levels of government, the media, academia and policy analysts 'who play important roles in the generation, dissemination, and evaluation of policy ideas' (Sabatier, 1988: p. 131).

The relevance of Sabatier's ideas to adaptation policymaking is noteworthy. The research presented in Chapters 3, 6 and 7 suggests that, in both Queensland and in the UK, the climate change adaptation policy 'subsystem' has competed with concurrent subsystems relating to disaster risk and natural resource management, amongst others. Moreover, both stable and dynamic socio-economic and political structures, alongside advocacy groups – similar to those coalitions described by Sabatier (1988) – appear to have had substantial influence upon climate adaptation problem framing and policymaking in these two countries.

Each of the contrasting ideas of incrementalism, bureaucratic politics, structuralism and social networks may have a partial validity, though perhaps none describe policymaking with adequate clarity on their own. As I seek to show here, the precise dynamics of the policy process are often dependent on the norms and socio-economic parameters and circumstances influencing policy players, and the available resources of a policy subsystem within any given context. What is important to note for the purposes of the arguments made in this book is that linear technocracy, or related ideals of scientific rationality under threat from political irrationality, are wholly insufficient models of the evidence-based policy process. As I discuss in later chapters, prevailing uncertainties, the limitations of presumed rationalist and incrementalist policy models, and the presence of political influence in both evidence development and bureaucratic decision-making, raise important questions about what constitutes legitimate knowledge and expertise for evidence-based adaptation.

What is evidence for policy? Science, appraisal and the coproduction of policy evidence

Despite rationalist assumptions to the contrary, evidence for policy should no longer be (if it ever really was) the sole responsibility of scientists, economists and other certified experts. As Head (2008a), Collins and Evans (2002) and Solesbury (2001) amongst many others have argued, our ideas about what constitutes policy evidence have changed dramatically since Price's (1965) invocation of science as a means to 'speak truth to power'. The science of the natural environment is not what we thought it was (i.e., the purveyor of immutable truths) and even if it was, there are other forms of knowledge required for making good policy decisions. In recent decades, despite a continuing adherence to the linear-technocratic heuristic, policy players have begun to take on board ideas about the necessarily partial nature of expert assessment for public decision-making, particularly those associated with the natural environment. On this basis, there have been increasing calls for the democratisation of expertise for policymaking[3] (Jasanoff, 2003a, 2005; Irwin, 2006), calling for broader concepts of what expertise is, greater public participation and oversight in evidence development, and greater access to expert knowledge for all policy players and the public.

Constructivist ideas about what constitutes valid evidence and experts' valid contributions in the public sphere have evolved gradually from the realisation, initially argued for instance by the likes of Karl Marx and Thomas Kuhn amongst others (Restivo, 1995), that scientific research and expert analysis, even when separated from broader society, are themselves types of social interaction and that problem framing and characterisation have a necessarily normative constituent that is not the sole preserve of technical experts (see for example, Jasanoff and Wynne, 1998; Douglas and Wildavsky, 1983). Moreover, technical decisions and the associated science needed to inform them often involve significant subjective and normative judgements which mean that scientific answers cannot legitimately be considered wholly impartial. Scientific investigation even when removed from its broader social context, is not and cannot be entirely value-free. Heather Douglas (2009; 2004), amongst other philosophers of science, has shown how values are an inevitable part of deriving robust science. She describes a broad distinction between *epistemic* values on the one hand, and *non-epistemic* socio-cultural values on the other, that can influence scientific practice.

Epistemic values are those incorporated in the methodological choices involved during the practice of science that are not relevant to broader society. Citing Churchman (1956, p. 248) Douglas (2009: p. 55) suggests that these values have traditionally been shielded from extrinsic political influence; they are circumscribed by the prevailing norms and best practice of a scientific community concerning:

1   the relative worth of further observations or empirical investigation when confirming the veracity of a hypothesis;
2   the scope of a scientist's conceptual model;
3   the value placed upon simplicity when assessing the relative worth of a hypothesis;
4   the need for precision of language in describing observations and results;
5   the accuracy of the probability assigned to the truthfulness of a hypothesis.

Non-epistemic values, by contrast, are those that relate to the application of science to the real world and are implicated in moral, political and social judgements that are relevant outside the scientific community's good methodological practice (Roberts, 2007). In later writings, Douglas (2009) divided epistemic and non-epistemic values into more specific sub-categories but it seems clear that non-epistemic values should have no legitimate role in good science. As I discuss in Chapter 4, however, the distinction between, and the valid roles for, epistemic and non-epistemic values and their variants is not entirely clear during the derivation of credible and usable climate and adaptation science for policy. This conceptual distinction is nonetheless useful for understanding the basis of my arguments about a politicisation-by-process that occurs in the development of science for climate adaptation (see Chapter 4), versus a politicisation-by-agency that may be more likely to occur during policy evidence development (discussed in Chapter 7). Both types of politicisation ultimately entail more than strictly epistemic values in the derivation of policy knowledge.

Beyond philosophical discussions about the value-content of science lie broader questions about the role of science in society that are also relevant to discussions about the interplay between expert and political authority for public decision-making. In this regard, social and political scientists often adhere to constructivist perspectives. The evolution and trajectory of constructivism has been widely discussed in recent years, in particular following a series of heated debates in the 1990s about the nature of science that became known as the 'Science Wars' (see Parsons, 2003 for an overview). This debate centred on whether facts were objectively real or simply constructed. As Sarewitz (2000, p. 80) points out, however, neither side of this disagreement spent much effort considering the ways in which both perspectives were (at least partially) correct: 'Facts are both objective (that is, representations of something real) and constructed (that is, products of social context)'. Constructivism in its weak form (see Chapter 1) originates from the viewpoint that science is not extrinsic to society, nor are scientists able to entirely distance themselves from their human frailties that make them prone to cognitive bias and logical fallacy. Weak constructivists do not question or reject the idea that objective truths exist, but simply that derivations and interpretations of what the facts are, which ones are important, and what they mean for decision-making, can vary depending on an expert's perspective.

One of the most influential contributions in helping to understand the constructivist framework and the role of values in science came from Weinberg (1972). Weinberg coined the term 'trans-science' to argue that science for policy often cannot or does not provide definitive answers and that judgements are usually required about the use and presentation of facts. Trans-science addresses questions that are asked by, or of, the scientific community, yet which require such significant subjective or normative interpretation that their attempted resolution for policymaking transcends the pursuits of objective observation and hypothesis-testing that characterise 'normal science' (Kuhn, 1962):

> I propose the term trans-scientific for these questions since, though they are, epistemologically speaking, questions of fact and can be stated in the language of science, they are unanswerable by science; they transcend science. In so far as public policy involves trans-scientific rather than scientific issues, the role of the scientist in contributing to the promulgation of such policy must be different from his role when the issues can be unambiguously answered by science.
>
> (Weinberg, 1972: p. 209)

The concept of trans-science not only highlights that scientific evidence for policy is not a homogeneous set of objective truths derived by the same basic means, it also helps to understand that value judgements are often necessary and inevitable when seeking to apply technical understandings to real-world decision-making.

As I describe in Chapter 4, when considering future climate change, normative choices during the derivation of evidence can make a significant difference to the conclusions we draw. For adaptation policy, trans-science seeks answers to questions such as 'What is normal climate?'; 'What will the climate be like in 2050?', or 'What will the impacts from a 2-degree Celsius rise in global temperature be?' Climate change involves such complex interactions between bio-geophysical and socio-economic systems, however, that experts' limited access to data and limited ability to understand the dynamics of these systems using the 'virtual science' (Kellow, 2009) of climate change models necessitates a range of subjective or normative assertions and interpretations, for which experts may not be able to claim privileged expertise. In other words, when applying their science to the real world, they may be obliged to make non-epistemic value judgements in order to derive robust and usable science for policy.

Jasanoff and Wynne (1998) have argued that the proliferation of scientific modelling in the last few decades has ensured that the answers to many trans-scientific questions sit securely within the remit of experts and the production of model output, even though many of the inputs and outputs of such modelling involve subjective interpretations of how the world is, or ought to be, about which the lay community may arguably have equally valid opinions. In a similar vein, but in a seemingly broader sense, Fischer (2009) suggests that expert evidence for policy requires both technical reasoning – as used in the pursuit of pure science – and a 'socio-cultural' reasoning, relating to the interpretation, evaluation and choice of norms and values that are an inevitable component in the development and choice of evidence for public policy. Fischer's ideas are discussed in more detail below.

Weinberg's trans-science highlights the limits of conventional ideals for reductionist science when informing adaptation policy. However it also alludes to the inadequacy of scientific expertise on its own when directly addressing complex, contentious, pluralistic and location-specific problems. Indeed, as the case-studies in this book show, trans-scientific endeavour cannot objectively resolve many of the value-based questions posed in the development of evidence for policy. In this way, Funtowicz and Ravetz (1993) proposed the concept of 'post normal science' to account for the relevance of viewpoints outside the field of conventional expertise in the provision of evidence, in circumstances where time is limited and uncertainties and decision stakes are high. As discussed by Jasanoff and Wynne (1998), post-normal science brings the spheres of science and policy much more closely together by arguing that valid technical interpretations of social problems cannot be separated from value commitments that underpin their assessment. As such, post-normal science must be produced by a range of relevant stakeholders: experts, bureaucrats, politicians and the public, who can and should contribute to the production of policy knowledge.

The concept of post-normal science has not gone without criticism. Wesselink and Hoppe (2011) suggest that post-normal science, although presented as a new way of doing science, may actually just be another form of politics through the democratisation of evidence-based policymaking. Collins and Evans (2002, p. 282) meanwhile argue that post-normal science is unhelpful because it treats different types of expertise and knowledge as interchangeable and therefore does not help to overcome the 'problem of extension': the tendency to broaden the concept of expertise too far, and thus for the boundary between expertise and politics to be blurred to such an extent that it is difficult to understand where expert authority begins and ends.[4]

While granting some of the criticism levelled at the post-normal thesis, I argue nonetheless that the term can provide a useful conceptual frame for understanding the ways in which lay knowledge has become increasingly important for understanding the context-specific technicalities of environmental problems (Juntti et al., 2009). In particular, as I describe in Chapter 4, the concepts of post-normal science and trans-science help to elucidate the characteristics of the differing components of scientific knowledge concerning climate change, the extent to which climate science can ever be considered wholly objective, and the extent to which policy is ever actually evidence-based in the ways expected by linear-technocratic models. Distinguishing between different forms of evidence and expertise through the use of these concepts also helps to understand why it is difficult to discern where exactly expert authority ends and political authority begins when considering the value of evidence for climate adaptation policy (Fischer, 2009; Hertin et al., 2009; Dryzek, 1993).

The evolution of ideas concerning the character and role of science and expertise in society has been summarised in a number of ways by those who, like myself, seek synopsis. For instance, Gibbons et al. (1994) described alternative concepts of science and research for public policy by comparing a traditionally assumed *Mode 1* knowledge production with a more contemporary and realistic *Mode 2*. Mode 1 conceptualised knowledge production as purely scientific, largely dictated by the outputs of theoretical and experimental science and governed by assumptions that scientists and their institutions dictated their own research agendas. Under Mode 1, science can and should speak truth to power, as per for example, the linear-technocratic model. By contrast, Mode 2 conceptualised knowledge production in the context of its application to decision-making. Whereas Mode 1 is homogeneous and situated within specific scientific disciplines, Mode 2 is necessarily transdisciplinary and heterogeneous. Under Mode 2, for contemporary policy issues – characterised by high levels of uncertainty and complexity – both expert and non-expert policy players and stakeholders can contribute relevant knowledge and should be involved in the development of policy evidence.

The concepts of post-normal science and Mode 2 knowledge production are also related to ideas concerning the coproduction of policy knowledge, a similar type of constructivist/ critical realist concept developed by the science and technology studies (STS) community:

> [C]o-production is shorthand for the proposition that the ways in which we know and represent the world (both nature and society) are inseparable from the ways in which we choose to live in it ... Scientific knowledge, in particular, is not a transcendent mirror of reality. It both embeds and is embedded in social practices, identities, norms, conventions, discourses, instruments and institutions.
>
> (Jasanoff, 2004: pp. 2, 3)

Coproduction does not argue for a completely relativist view of science, but rather should be viewed as a critique of realist ideology and positivist methods of evidence development that would envisage nature, facts, reason and policy as separate from culture, values and politics (Jasanoff, 2004). Under the coproduction model, science is a social endeavour and another component of a broader society. In this book, I envisage knowledge coproduction in rather less abstract terms, and use the concept as a means to bridge the epistemic interface between science and politics. In Chapter 4, I propose the term 'coproduced knowledge' to describe the process of bureaucratic evidence development that strives (somewhat unsuccessfully at the current time) toward a balance between the technical credibility and political acceptability of policy knowledge.

Mimicking the terminology of Gibbons et al. (1994), Collins and Evans (2002) influential paper summarises these various related perspectives by suggesting that science studies has undergone three waves of academic development. The first wave aligned with positivist ideals and Mode 1 science that was incontrovertible and could apprehend objective reality. The second wave sought to address what Collins and Evans (2002: 237) called the 'Problem of Legitimacy' that arises because scientists cannot justifiably claim legitimate privileged access to a universal truth in many matters relating to science and society. Scientific endeavour that seeks to understand and ameliorate society's relationship to the natural world is itself a social process of knowledge construction so that empirically valid facts may not be wholly accurate representations of objective reality. The second wave therefore showed that 'the basis of technical decision-making can and should be widened beyond the core of certified experts' (Collins and Evans, 2002: p. 237), thus appearing to resolve the problem of legitimacy. In its place, however, a 'Problem of Extension' arose, by which it is difficult to know where to draw the bounds of expert authority for rationalising democratic policymaking.

These various theoretical conceptions of technical evidence and expertise suggest that the privileged position of science for policymaking has been overstated in the past, and yet, science is still clearly a necessary and desirable contribution for informing policy debates given the enormity of its contribution to contemporary society. There are other forms of evidence besides the natural sciences, however, that are increasingly relied upon to inform policy (Head, 2008b; Hertin et al., 2009; Solesbury, 2001). Social-scientific research in particular has been used since the resurgence of evidence-based policymaking in the late 1990s. Applied social sciences have been increasingly influential for understanding public policy issues, and amid increasing calls for the democratisation of evidence-based approaches that conceive of expertise in broader terms, accounting for multiple perspectives and forms of valid knowledge (Irwin, 2006; Maasen and Weingart, 2005; Solesbury, 2001). The role of social science will be discussed in Chapter 7 in the context of understanding risk-based approaches to adaptation policymaking.

Alongside the natural sciences and social-scientific research, *ex ante* policy assessment and appraisal have also been increasingly used by liberal democratic governments that pledge allegiance to the evidence-based mandate. *Ex ante* evidence development for policy attempts to more effectively bridge the divide between science and policymaking (Nilsson et al., 2008; Hertin et al., 2009; Owens et al., 2004). Technical *ex ante* assessments, such as risk and impact assessments, are the subject of contrasting views in the academic literature in terms of their epistemological status and whether they do in fact sit at the science-policy divide. Some, particularly in the US, have suggested that these assessments can be the principal product of expertise, distinct from the subsequent task of risk management which addresses the politics of risk (see for example, Keller, 2009: p. 12). Alternatively, Jaeger et al. (1998) and Owens et al. (2004) suggest that technical *ex ante* assessments have an inherent and unavoidable political component. Owens et al. (2004: p. 1943) define *policy appraisal* as: '*ex ante* techniques and procedures that seek to predict and evaluate the consequences of certain human actions'. Examples of commonly used forms include environmental impact assessment (EIA), strategic environmental assessment (SEA) and sustainability appraisal (SA) used for the purposes of urban and regional planning. Other scholars discuss *ex ante* assessment in broader yet similar terms that include cost benefit analysis and risk assessment alongside less formal heuristics, checklists and decision trees (Turnpenny et al., 2009; Radaelli and Meuwese, 2010; Hertin et al., 2009). Importantly, this latter group of scholars often also conceptualise *ex ante* policy appraisal as undertaken (rather than merely overseen) *by* bureaucrats *for* policymakers, sometimes even without the direct involvement of technical experts.

Ex ante policy assessment nonetheless often seeks to bridge (if not in actuality, then by rationalist pretence) the science-policy divide through the provision of salient technical knowledge to inform and justify policymaking. On a political view of bureaucratic policymaking, these assessments would involve processes of problem definition, framing and prioritisation that are inevitably subject to influence from the values and priorities of those undertaking such assessment. Thus, they may be delivered and framed in ways that ensure a preferred course of action is taken (Owens et al., 2004: p. 1946):

> The use of appraisal techniques as "post demonstrations of preconceived judgements" has been widely documented ... More insidiously, techniques that are ostensibly neutral may in fact have an in-built tendency to support particular outcomes ... ethical and political choices masquerade as technical judgements, reinforcing prevailing norms and existing structures of power.

In summary, the various literatures discussed here identify a number of bodies of knowledge as suitable evidence for policymaking: natural science (including trans-scientific constituents that require experts to make significant value judgements, and post-normal scientific processes requiring non-expert input), social science (including economic appraisal), the local and contextual perspectives and knowledge of relevant stakeholders, as well as the outputs of *ex ante* policy assessment such as risk and impact assessments that may incorporate all of the above. Although the linear-technocratic model may consider these forms of evidence in a hierarchical fashion that prioritises supposedly objective quantitative science over all else (Marston and Watts, 2003), the manner in which climate science and other evidence is developed and used suggests that policy knowledge for climate adaptation must be considered much more than just objective technical research and assessment or the impartial and linear interpretation of science. As alluded to in Chapter 1, in this book I also propose a distinction between the general concepts of evidence as a broad body of available knowledge, versus a more specific idea of *policy evidence.*

Evidence, broadly described, can be considered to be the outputs of expertise and accumulated public knowledge developed and used under the scrutiny of peer-review. This concept aligns with both the outputs of the scientific and social-scientific communities (those traditionally assumed to hold expertise), as well as with second wave ideas relating to the democratisation of expertise (Maasen and Weingart, 2005), whereby local and contextual knowledge may constitute valid forms of non-certified expert evidence under the peer-review of various interested parties in the development of public knowledge and decision-making (Collins and Evans, 2002). By contrast, I propose that *policy evidence* may be considered a sub-set of this broader body of knowledge, developed specifically for the purposes of policy decision-making. Policy evidence is often summarised and collated in the form of *ex ante* policy assessments, and may use various forms of expertise and knowledge, including those from the academy; from those versed in local and contextual perspectives; as well as the views of policymakers themselves who claim expertise over the socio-cultural context of developing policy subsystems. What distinguishes policy evidence from the broader body of evidence is that its development is overseen by bureaucrats for policymakers. As I discuss in detail in later chapters, policy evidence defined in this way appears to necessitate political as well as expert input and is developed and used in many more ways than described by the linear-technocratic schema to which this knowledge nonetheless often subscribes.

The role of policy players in evidence development

The role of bureaucrats

I hope at this point that I have conveyed the prevalence, if not also the potential legitimacy, of non-expert perspectives in the development of policy evidence. Trans-science, post-normal science and the concept of 'wicked' policy problems highlight how certified experts struggle to justify the legitimacy of their sole *privileged* authority over technical decision-making in the public realm. Such decision-making often requires – implicitly or otherwise – socio-cultural, ethical and normative value judgements that lie within no one's precise expertise. The potential for political influence in bureaucratic policymaking described at the start of this chapter, therefore, is also important for understanding policy evidence development in light of the complex, uncertain and contentious nature of contemporary issues like climate change (Head, 2008a). As I will describe in later chapters, however, the impetus for political influence in the development and use of evidence for policy may be enhanced by expectations for both the democratisation of expertise (Irwin, 2006), as well as by the inevitable limitations of reductionist science when attempting to fulfil the role expected of it by linear-technocratic models of decision-making (Keller, 2009). It is important, therefore, to consider the role, not just of experts, but of non-expert policy players involved in the development and use of evidence.

Since the Enlightenment era of the eighteenth century or so, there have been recurring expectations amongst government and the public that evidence and expertise should be used to ensure that policymaking is pursued through objective, systematic processes of accountability (Kitcher, 2011) to ensure that it is rational. This expectation was perhaps first institutionalised in nineteenth century Britain, one of the pioneering nations in the development of a supposedly neutral expert bureaucracy, designed to implement the wishes of the political executive (Bayley, 2004; Gerth and Mills, 1970). In theory, the role of bureaucratic institutions was envisaged as an organisation of both expert policy advisers, as well as being responsible for the administration and implementation of policy directives coming from elected government (Parsons, 1995).

Max Weber was one of the first to describe the role of bureaucracy in his 'rational control' model of bureaucracy. This model envisaged policymaking in a top-down fashion; a discrete task completed by elected government executives (Parsons, 1995: pp. 463–465). These executives are then advised by, and pass decisions on to, neutral (i.e., disinterested), objective bureaucrats who administer complete policies. This simplistic model assumed an effective and clearly defined chain of command, a capacity to co-ordinate and control within governance systems, and perfect communication in and between units of government organisation. It placed considerable emphasis on policymaking at the top of government hierarchy. Of course, few would argue nowadays with the idea that, in reality, bureaucracies also hold considerable policymaking power (du Gay, 2005; Page, 1985). As discussed, for instance, by Lindblom (1959) policymaking does not come to an end once a policy is set out or approved by government (Peters, 2003; Parsons, 1995: p. 465; Page, 1985).

In practice, bureaucrats are not neutral servants of the state. They have ideas, values, beliefs and interests which also shape policy. Indeed Peters (2003) and Page (1985: p. 32) argue that Weber's view of bureaucracy was not as apolitical as it is often depicted and that he was also aware that bargaining over values was inherent in the public sector and that policy was also made by public servants. Although the forms and interpretations of the role of bureaucracy have changed considerably from the Weberian ideal, in particular since the 1980s as a result of neo-liberal reforms that became known as New Public Management[5] (NPM) (Head, 2008b; Bevir et al., 2003), bureaucratic institutions still fulfil key government functions in relation to overseeing the development and collection of policy-relevant evidence (Peters, 2003; Bevir et al., 2003; Gerth and Mills, 1970: p. 196). And despite numerous predictions to the contrary, bureaucratic structures have yet to be replaced by other forms of policy administration (du Gay, 2005: p. 1). Thus, bureaucracies may be considered to be, in many ways, at the interface between political and expert authority, even though their original conceptualisation was as a source of expertise itself (Hoppe, 2005; Bruner and Steelman, 2005: p. 23; Leiss, 2000: p. 50). Dye (2005: p. 52) agrees that policy implementation is the continuation of politics by other means. He argues that although bureaucracy is not constitutionally empowered to decide policy questions, it does so nonetheless, as it performs its task of implementing the wishes of the political executive.

Maasen and Weingart (2005) make an important point in this regard, highly relevant to evidence-based climate adaptation and climate change policy more generally. Whereas the modes of communication and investigation through science are designed toward questioning existing knowledge and reducing uncertainty in order to advance evidence-based decisions, the motivations of policymakers, by contrast, are aimed at closing down public dispute through consensus building and compromise, using knowledge as required. In political debate and bureaucratic policymaking, therefore, it is convenient if expert knowledge supports consensus or compromise and very inconvenient when it conflicts with and therefore de-legitimises past or future decisions or present consensus. Thus, the continued production of science continually irritates politics in unpredictable ways and ensures that bureaucratic policy players are incentivised to engage with the production of policy evidence.

Given the uncertain, complex and contentious nature of climate adaptation problems (see Chapter 4), therefore, the role of bureaucrats as policy mediators, advisers, designers and administrators takes on particular significance when considering evidence-based policy. As I seek to show in this book, bureaucrats can be intimately involved in decisions concerning how we should understand the risks and benefits associated with climate change, how they should be portrayed to ensure their political acceptability under an evidence-based mandate, and even, what constitute the 'facts' in the first place. These ideas concerning the role of bureaucracy also add weight to the arguments of Schon and Rein (1994), Lyles and Thomas (1988) and Rittel and Webber (1973) (see Chapter 4) that the way governments choose to define and characterise a problem dictates how they attempt to resolve it, and ultimately influences the effectiveness of their response. The case-studies examined in this book demonstrate how bureaucratic government is not just involved in the framing of scientific conclusions but is also a key contributor to the development and/or pre-selection of those conclusions in a way that suggests considerable overlap between the realms of expertise and politics.

The role of scientists and other experts in policymaking and evidence

Although various social science disciplines have described numerous ways in which evidence may inform policymaking; there seems to be considerably less concern about the specific roles for scientists and other experts in policymaking. Indeed it is often difficult to speak about experts in policymaking without also discussing the nature of the evidence they wield. Perhaps a good starting point in this regard is in the work of Roger Pielke Jr (2007) who discusses four idealised expert roles:

1   **Pure scientist:** this role views the expert as providing discrete bodies of facts in the course of his/her pursuit of knowledge, and nothing else. The Pure Scientist has no direct interaction with the decision-maker and assumes that what the latter subsequently does with that information lies beyond the remit of scientific advice.
2   **Science arbiter:** in this role the expert waits at the behest of the decision-maker ready to provide objective facts and impartial expert judgement that can be resolved with science alone as required. The arbiter avoids getting involved in non-scientific issues or normative/political questions.
3   **Issue advocate:** this role suggests that the expert uses his/her expertise to advise the decision-maker in favour of a particular course of action, based on the expert's values and normative interpretations of the facts.
4   **Honest broker:** the expert provides all the facts relevant to the decision-maker's choice and attempts to integrate the available evidence into the policymaking sphere. The honest broker seeks to clarify the scope of policy choice in a way that helps the decision-maker, in line with their preferences and values and considering other stakeholders' views.

Spruijt et al. (2014) have taken a considerable step in clarifying these roles, demonstrating through a cross-disciplinary literature review how, despite a general scarcity of supporting empirical evidence, there is general agreement across social science disciplines that the role of experts depends on the type of problem being addressed (simple or complex; wicked versus tame), and the influence of other parties (bureaucrats, other policy players and the public).

As discussed above, many scholars stress the apparent myopia of policymakers concerning the nature of expertise and evidence for policymaking based on positivist assumptions about what science and other evidence can do. They advocate for a better understanding amongst policy players about the inability of scientific expertise to unequivocally or objectively understand the technical aspects of policy issues (Nowotny, 2003; Turner, 2001; Jasanoff, 1990, 2003b). Some propose that knowledge production move away from concepts of reliable expert knowledge toward, instead, socially robust knowledge, embracing the aforementioned ideas concerning the democratisation of expertise (Nowotny, 2003) and the need for legitimate and salient policy evidence (Cash et al., 2003).

These ideas also align with Beck's (1992) seminal sociological thesis, in which he argues that although scientific understanding has brought undoubted benefits to society, its use is also the cause of many problems, including a need to contain and manage risks endemic to its utilisation in modern society. Under this view, scientific endeavour is both the cause of and proposed solution to many contemporary problems associated with modernity and has done much to increase uncertainties for decision-makers. These difficulties present significant challenges for understanding the valid roles for scientists in the practice of policymaking.

Weingart (1999) and Sarewitz (2004) illustrate how contrasting norms and values are a recurring component in both the development and use of policy evidence for problems like climate change. This is not just because of the many necessarily value-based, normative choices required when developing useful technical evidence (discussed in relation to trans-science and post-normal science above) but also due to how multiple disciplines of scientific expertise can hold contrasting, sometimes conflicting, but equally valid knowledge about the natural world. These understandings may therefore align with numerous contrasting normative viewpoints about highly complex and uncertain systems, so that there are multiple valid and potentially contradictory 'facts' about the world which compete for attention in the spheres of policymaking and politics.[6] These ideas complicate the simplified ideals for expertise provided by Pielke Jr (2007) above, by suggesting that facts and scientific knowledge are rather more mutable when it comes to developing policy evidence than we often assume (Juntti et al., 2009; Sarewitz, 2004). Spruijt et al. (2014) suggest that in order to ensure the acceptability of policy measures, therefore, it is necessary to acknowledge the validity of many potentially contradictory expert viewpoints about what constitutes a risk, or what counts as valid 'propositional knowledge' concerning what is a problem and why (Collins and Evans, 2002).

To further complicate these problems of deriving a single set of valid expert knowledge, the questions policymakers ask of scientists are often not answerable within the strict parameters of their precise expertise (Fischer, 2009; Jasanoff, 2005). Instead, experts are expected to act as persons with analytical skills grounded in practice and experience, rather than those with 'unmediated access to ascertainable facts' (Jasanoff, 2005: p. 211). Although they are expected to have mastery of a particular area of knowledge, they are also expected to be able to size up bodies of heterogeneous knowledge and to offer opinions based on imperfect understanding, on issues that lie within nobody's precise disciplinary competence. These difficulties appear to limit experts' role in fulfilling Pielke Jr's (2007) idealised participation in policy advice as apolitical *pure scientist* or *science arbiter* when directly informing policymaking. The role for experts as the purveyors of impartial knowledge and enlightened judgement, therefore, also raises important questions about their supposed neutrality, and, the neutrality of liberal democratic policymaking more generally. In this regard, Turner (2001, p. 215) pointedly asks: 'if the liberal state is supposed to be ideologically neutral, how is it to decide what is and is not ideology as distinct from knowledge?'

Jasanoff (2005) argues that although the practice of science for policy has attempted to evolve in order to integrate with the political realities of policymaking, the rhetoric around policy science has not caught up with the realisation of the partiality of expert knowledge and expert evidence. The general call is still for objective, untainted knowledge as a means of justifying policy decisions. Conversely, there seems an ongoing need for greater public participation in the development of policy knowledge, both as a means of generating knowledge itself and as a means to 'facilitate learning and mutual understanding and prevent unnecessary conflict' (Spruijt et al., 2014: p. 21). Maasen & Weingart (2005) and Irwin (2006) suggest an interesting paradox in this regard. On the one hand, the evidence-based mandate has called for, and largely received, a large increase in the production and use of supposedly impartial expert evidence for policy. On the other, it has been accompanied by a commitment from policymakers for greater inclusivity of stakeholders and the public in evidence development, speaking to the democratisation of expertise.

Irwin (2006: pp. 301,302) suggests that policy documents from the UK and the EU are prone to contradicting themselves through their rhetoric, whereby: '[A] dominant "inclusive" voice stresses public dialogue, and a second, "scientistic" voice tells the reader that the public can only make its contributions if it is properly instructed and educated.' Thus, policymakers appear to be seeking to partially substitute a previously assumed deficit of understanding (Bucchi and Neresini, 2008) with a deficit of public trust in the available science in order to facilitate policymaking, while simultaneously maintaining a rhetoric advocating for the value of impartial and robust scientific expertise.

Irwin (2006) suggests that we should be cautious of such commitments to inclusivity as a result, and that we need new epistemological and political understandings of expertise in order to change the science–public relationship in a way that can meaningfully achieve its democratisation while maintaining its decision-making authority. I agree. The available literature suggests that the role of experts is far from clearly delineated, and indeed that, at best, policymakers have held to the notion of objective, value-free expertise in the absence of a more useful model of evidence-based policymaking. In any case, as I argue here, this pretence is politically expedient for protecting both experts' (as the purveyors of objectivity) and policymakers' (as rational decision-makers) legitimate roles (Owens et al., 2004; Cash et al., 2002).

Despite the persistence of rationalist approaches to evidence use and policymaking, Maasen and Weingart (2005) point to some noteworthy trends in the pursuit of the evidence-based mandate. In particular, the tendency for partisan expertise and knowledge to be used and provided by think-tanks advising government under NPM structures. Expertise has increasingly become a commodity whereby both public and private institutions produce knowledge irrespective of demand in order to demonstrate their expert credentials should their advice be sought:

> As expert knowledge has grown in importance as a political resource, actors in the political arena attempt to obtain and control the knowledge that is relevant to their objectives. This competition for knowledge, which

already represents 'democratization by default', has resulted in the loss of science's monopoly on pronouncing truths.

(Maasen and Weingart, 2005, p. 7)

Bureaucracy, it seems, is capable of picking and choosing which knowledge to use, even though they may have increasingly less responsibility for its production than they once had. This trend suggests a diminishing space for Pielke Jr's *honest broker* than might at first seem to be the case. Since there are so many experts and so much public knowledge to choose from, expert authority may not be chosen on the basis of credentials or qualifications but on the degree of congruence of experts' views with the values and goals of decision-makers. Would-be honest brokers, therefore, must compete with other experts for their voice to be heard.

Given the increasingly uncertain, ambiguous and incomplete nature of expert knowledge for policy, the remaining persona of expertise described by Pielke Jr (2007), the *issue advocate*, becomes increasingly tenable; one that speaks to the direct politicisation of expertise within the market place of ideas. Indeed, renowned postmodernist scholars such as Foucault might contend that distinctions between expert and political authority are largely nonsensical anyway, that there is little to distinguish an expert advocate from a political one (Turner, 2001). I fear that the Foucauldian perspective is more than a little extreme but, in later chapters I nonetheless highlight the importance of realising and understanding the potential roles for expert advocacy and politicisation in climate adaptation policymaking.

Issues of demarcation at the science-policy interface

In response to the various arguments discussed above, scientists have often been sceptical or ambivalent, if not also hostile toward stronger constructivist forms (see for example, Gross and Levitt, 1998). Such disapproval is in no way surprising from those of a positivist persuasion who tend to assume that social constructivism entirely denies the empirical validity of contemporary scientific facts; and to be fair, some social constructivists appeared to do just that (see, for example, Latour and Woolgar, 1979). More sophisticated commentary concerning the relative worth of scientific positivism and social constructivism, however, has noted that although scientific knowledge may indeed be influenced by social factors, it would be foolhardy to assume that science and scientific expertise was not still a special form of knowledge to be embraced for public decision-making. In reference to the second wave of science studies that has occurred since the 1970s (see above), and which sought to understand the social contingencies of scientific expertise, Harry Collins and Robert Evans (2002: p. 239) noted:

[S]ociologists have become uncertain about how to speak about what makes [scientific knowledge] different; in much the same way, they have become unable to distinguish between experts and non-experts. Sociologists have become so successful at dissolving dichotomies and classes that they no longer dare to construct them.

If we accept the weak constructivist stance then it seems that, since those who develop science and policy evidence will inevitably be influenced by the social and political norms and contexts within which they work when deciding which science is relevant and what expert judgements to make, questions concerning what constitutes *valid science* for adaptation policy are inevitably intertwined with what constitutes *valid expertise*. Both evidence and expert legitimacy in this regard are relevant to the ideas in this book, particularly since climate adaptation science and policy evidence are often developed through coproduction processes between certified and non-certified experts, as well as with non-expert policy players. A challenge, therefore, lies in deciding who counts as an expert for adaptation problems and what should count as expert evidence.

As I explain in detail in Chapter 4, establishing universal truths concerning future climate change and, thereby, irrefutable expert knowledge about adaptation problems, is no easy feat. Furthermore, demarcation criteria associated with commonly held norms of 'good science', such as falsifiability, disinterestedness, communalism, universalism and organised scepticism (Gieryn, 1995), that would allow policymakers to rely primarily upon science and scientific judgement to effectively adapt, are limited by their poor applicability to many technical problems of public policy. The science and expertise associated with climate change in particular often struggles to fulfil these criteria (see for example, Hulme (2009), Kellow (2009) and Oreskes (1994)). This is due to a few reasons. To begin, the science of *future* climate change depends on subjective and normative assertions about the trajectory of a complex dynamic open system, humans' agency within this system, and over temporal and geographic scales that preclude scientific certainty for policymaking. The adaptation policy problems to which this science and expertise might be applied, on the other hand, tend to be context-dependent, time-limited (i.e. decisions are often made over short timescales) and hyper-politicised. Under these conditions, decision-makers are unable to reliably depend upon conjectures (e.g., humans cause *dangerous* climate change) that are only falsifiable in a *post hoc* manner. Decision-makers must therefore depend upon scientists' generalised judgements based on their climate and adaptation science, about a suite of context-specific and often highly interconnected and complex risks to society, economy and ecology, that no one can claim wholly privileged expertise over (many others may hold valid knowledge about these problems) and that are, therefore, vulnerable to the strongly partisan politics of climate change.

Scholars such as Gieryn (1995) and Guston (2001) argue that debates concerning what demarcates science from non-science when informing public decisions have been characterised by two contrasting fields of practice:

essentialism and constructivism. Essentialists are pragmatic in the sense that they develop demarcation criteria in the course of their practice of using science for decision-making – what Gieryn (1983, 1995) calls 'boundary work' – in ways that constructivists might consider to be subjective, value-based, ambiguous and contextually contingent. Constructivists, accordingly, critique those demarcations made by essentialists and argue that there are no universally valid demarcation criteria; what constitutes *legitimate* science and expertise for decision-making depends on the decision-making context at hand.

Earlier in this chapter I mentioned the 'problem of legitimacy', described by Collins and Evans (2002), which tackles this very problem: trying to distinguish what constitutes acceptable and privileged expertise for technical decision-making in the public sphere. Collins and Evans (2002) describe how a third wave of science studies could address the resulting problem of extension: if certified experts do not have privileged access to universal and immutable truths about *public* problems, as revealed by the second wave of science studies, then how far should the democratisation of expertise extend? Just because science is subject to normative influence, does this mean that everyone should have an equal say in answering technical questions in the public realm? Collins and Evans' (2002) suggest not. Their influential ideas address the issue of demarcation between expertise and politics for technical decisions by suggesting that experts' role should be bounded on the basis of their practical experience with any given problem.

By attempting what they called a 'normative theory of contributory expertise', they sought to address the joint problems of 'legitimacy' and 'extension' by suggesting that an expert's decision-influencing authority should be granted on the basis of their contribution to the field of knowledge in question; valid expertise should be circumscribed by an actor's experience in dealing with the technical questions for which decisions are needed. Though we may sometimes struggle to justify experts' claims to privileged knowledge (e.g., where they make non-epistemic value judgements in the course of doing science and developing evidence), or to precisely locate the boundary between expert and political authority for public decision-making, this does not mean that the dichotomy between these realms of influence should be dissolved entirely.

Collins and Evans' (2002) paper prompted a volley of counter-argument by eminent Science & Technology Studies scholars such as Jasanoff (2003c), Wynne (2003), Rip (2003) and Fischer (2009). Much of their protest centred on Collins' and Evans' imposition of an erroneous idea that concepts of expertise (who knows what) and epistemology (how and therefore what we validly know) should be understood by everyone, and in every context, in the same way. Although scientists have traditionally established their expert authority on the basis of their ability to derive empirically valid truths under agreed methodological controls, this does not necessarily grant them equivalent expertise for technical decisions in the public realm. Public problems requiring technical decisions relating to, for instance, climate and the natural environment, often require the development of a natural history[7] and not solely or simply inductive inferences (Rip, 2003). In the development of such a natural history, local and contextual knowledge may have equal and even surpassing validity, and deciding who counts as a legitimate expert therefore depends on what types of technical problem are in question and the social, historical and institutional contexts in which they arise (Jasanoff, 2003c; Rip, 2003; Wynne, 2003). The public should therefore have oversight of the framing of issues that experts are asked to resolve (Jasanoff, 2003c).

Despite their subsequent assertions to the contrary (Collins and Evans, 2003), critics claimed that Collins and Evans' had incorrectly assumed that all technical policy issues (should) derive their core meaning from science; that science conducted under laboratory controls is the same as science conducted and applied in the wider world; and therefore, that issues concerning the legitimacy of scientific knowledge in the public sphere are only about propositional questions of fact and not also about *which* questions and answers are salient in the first place. In doing so, Collins and Evans were accused of 'reinforcing an illiberal cultural imagination based on uncritical acceptance of western scientism' (Wynne, 2003: p. 402). In light of this argument, one might wonder, if contributory expertise were used as a demarcation criterion for understanding climate, how might we apply it in the context of using evidence and expertise for adaptation policymaking?

Collins and Evans were accused of implicitly arguing that experts should be allowed to extend their privileged authority over the *framing* of technical questions and the meaning of associated problems (e.g., to what extent climate change should be perceived as dangerous), alongside providing answers to those propositional (factual) questions in the first place (e.g., that there is a significant climate change phenomenon and if so, what the associated hazards are). The former task, critics argued, is situated within no one's expert purview and depends largely upon context and perspective. Climate change might seem quite appealing to someone living in Siberia, for instance, and perhaps not at all 'dangerous' to someone who believes in an impending Judgment Day or the second coming of Jesus Christ.

Likewise, and as I argue in Chapter 3, the extant climate extremes and socio-economic interdependencies that have long prevailed in Queensland appear to have had an important neutralising effect upon political perceptions about the risks from, and therefore the need to adapt to, anthropogenic climate change; at least, any more than Queenslanders already need to be resilient to climate variability and extremes. A significant point of contention in this demarcation debate as it applies to climate change, appears to be to define which types of questions experts should have privileged decision-influencing authority over. If, as many have argued, climate is as much a cultural construct as it is a statistical aggregation of meteorological data (for a recent example, see Hulme, 2009), then why should scientists have sole authority in determining what is normal climate, what is dangerous, and how to understand climate change risks?

Critics of Collins' and Evans' also suggested an ambiguity in their thesis when explaining what exactly they meant by *technical* decision-making for public policy and the circumstances under which such decisions may be overseen by experts. Fischer (2009), in particular, argues that as soon as scientific experts extend their expert gaze beyond the 'hard sciences', a different perspective is required. The core esoteric scientific pursuits of investigation, observation, data collection and analysis can be, if not entirely objective in their execution, at least circumscribed by the established methodological and epistemological norms of the academy (Douglas, 2009). Such isolated scientific activity is therefore, to some extent, shielded from extrinsic political influence and in this pursuit practical experience might indeed be a valid basis for holding privileged status as a contributory expert. Fischer argues, however, that as soon as scientists begin to apply technical knowledge to questions concerning what is a public problem, why, and how it may be resolved, then a different form of rationality also comes into play.

Fischer (2009) argues that a 'socio-cultural' reasoning, alongside technical reasoning, is required for answering many seemingly technical questions in the public sphere. Even though many relevant decisions may still appear squarely within the remit and practice of science, the value judgements involved in socio-cultural reasoning cannot be claimed as the privileged domain of scientific expertise alone. As a result, the qualifications for participation in technical decision-making could often legitimately be extended to a broader range of stakeholders, rather than solely on the basis of experience with the science of the problem in question. Whereas Collins and Evans sought to demarcate scientific from political decisions on the basis of contributory expert knowledge, Fischer points out that even seemingly technical decisions often require non-scientific reasoning that contributory experts cannot claim privileged authority over. The importance of socio-cultural reasoning for Fischer seems to include but transcend the arguments made above concerning the importance of public determination of the salience of propositional questions and answers, and about the trans-scientific character of propositional questions for policymaking.

And so it is, I argue, with the science of climate adaptation. As described in Chapter 4, even questions of fact (and not only questions about *which* facts are important) must often be determined through negotiation between certified and non-certified experts and non-experts and are dependent on local and political contexts for their veracity and usability for climate adaptation. The idea, therefore, that policymakers should uncritically follow the advice of climate scientists and other experts, becomes difficult to justify.

Meanwhile, Darrin Durant (2011) argues that Collins' and Evans' ideas are simply concurrent with an alternative model of liberal democratic political philosophy. Constructivist arguments about the contingency of expertise presuppose that political deliberation and decision-making are more or less the same kind of process of reasoning within a liberal democracy. Durant argues, however, that Collins and Evans ideas align with those of political philosopher John Rawls about a separation of kind, between processes of democratic deliberation and democratic decision-making. Under this model experts should have a legitimate privileged role in technical decisions in the public realm, irrespective of the contingency of their expertise, because someone must make these decisions and why not let it be those identified as experts, practised in the assessment of evidence and use of logic and reason? Public deliberation (which should be egalitarian and for which expertise would have no (or limited) privileged status), under the Rawlsian ideal, is a separate activity from democratic decision-making where the privileged position of experts can be justified on the basis of political expediency. The aforementioned critics of Collins and Evans, however, hold liberal democratic ideals more in line with philosopher Jurgen Habermas' ideas that contingent expert perspectives can illegitimately colonise public meanings and therefore should not be allowed to make generalisable conclusions[8] that determine technical decisions in the public sphere (Durant, 2011).

Although adaptation scholars and practitioners may not often consider ideas concerning the socio-cultural content of technical decisions, my hope is that in the context of understanding the epistemology and politics of climate change science and the characteristics of climate adaptation problems, even natural scientists might concede that a strict positivist perspective is unhelpful for addressing *policy problems*. Technical decisions for climate adaptation resist the types of definitive or reductionist characterisation assumed to be available by positivists on the basis of laboratory or theoretical practices. As I seek to show here, common assumptions concerning objective expert authority, much like those relating to impartial apolitical evidence, can actually foment the politicisation of evidence-based policy for hyper-politicised problems like climate change. The epistemology of climate adaptation problems, coproduced evidence and expertise (see Chapter 4) and the revealed practices of bureaucratic politics and policymaking described in the case-studies herein (see Chapter 6 and 7), seem to make the clear identification of the bounds of contributory expertise, or the bounds of their privileged role in decision-making, challenging at best. Nonetheless, and as I discuss in Chapter 9, governments may indeed use the promise of apolitical expertise (though not necessarily that expertise itself) to expedite prevailing political priorities in a similar way to Rawls' ideal of a partitioned democratic decision-making process.

Even though I reject the positivist perspective and believe that the public should have oversight of the development of policy evidence, I also believe that social constructivists have, in the past, understated the value of technical expertise in this regard. Some constructivist scholars appear guilty of trivialising the empirical validity promised (and in so many other contexts, delivered) by science, as well as the systematic characterisation offered through scientific practice, even outside the laboratory. As a result, they underestimate the value of the *mechanistic* understanding and critical reasoning that science and scientific expertise provides, even in the face of uncertainty and potential politicisation.[9] On this basis, I believe that we should accept Western scientism *in a limited way* (i.e. as a valuable but insufficient cultural lens holding decision-*influencing* ability; see Collins and Evans, 2003) and concede that science is still pretty special relative to other forms of public knowledge.

The key point I wish to highlight by discussing problems of demarcation, however, is simply that although we may be able to distinguish a dichotomy between the realms of science and policy for climate adaptation, this does not necessarily mean that actors from one realm do not partake in the other during public decision-making. Nor does the presence of this dichotomy suggest that we can clearly identify where one realm ends and another begins. What is more important, perhaps, is the extent to which the value positions and judgements of both experts and non-experts are transparent during the development and use of evidence for policy. There is much validity to ideas of the historical and social contingency of expertise and the socio-cultural constituents of science in the public sphere, particularly for problems associated with future climate change. If nothing else, these social constructivist arguments highlight the potential for processes of politicisation and strengthen the case for oversight by the public upon scientific analysis for, and expert influence on, public decision-making.

Conclusion

Useful evidence for policymaking is at once both desirable – to ensure consistent and robust decisions – and a potential source of ambiguity and conflict for those who would seek to use it, due to the difficulties of drawing clear boundaries between expert authority and political influence. These difficulties are compounded when seeking to address wicked problems like climate adaptation. Not only are the bounds between expertise and politics difficult to draw, since the science used is inevitably influenced by normative choice and reasoning when used in the public sphere; but the very nature of the problem means that the facts themselves are often dependent on context, perspective and a wide array of intractable uncertainties that we can never hope to resolve, at least not to the satisfaction of policymakers within the timescales relevant for political decision-making.

This chapter has provided a theoretical foundation for understanding many of the difficulties associated with developing and using evidence for climate adaptation policy. Existing theoretical perspectives suggest that politics may be an inevitable component in the development of policy and therefore of policy evidence, particularly given the complex, uncertain and contentious nature of the problems this evidence seeks to address. As I demonstrate in later chapters, what constitutes valid adaptation evidence is not necessarily clear and is often dependent on where one sits at the interface between science and policymaking, or dependent upon the perspective of a particular level of governance. Moreover, how that evidence is subsequently used may be far from the ideal espoused by policymakers under an evidence-based mandate, or as described in the various policymaking guidelines and procedures devised by bureaucracies of contemporary liberal democracy. Perhaps a useful starting point for understanding the context-dependent nature of policy evidence and expertise, and the tensions between expert and political authority, therefore, is to understand the real-world contexts and perceptions of policy players in relation to the usefulness and usability of this evidence. In Chapter 3 I begin exploring this context by describing the political history and socio-economies influencing climate adaptation policymaking in the comparative cases of Queensland, Australia and England in the UK.

Notes

1 This observation, although broadly applicable for liberal democracies globally, seems partially incongruous with the findings of the research presented here, given Australian governments' apparent reticence toward science and the evidence-based mandate in recent years (Milman, 2014; Towell et al., 2013).

2 One of the most significant debates concerning policymaking during the twentieth century related to the viability of the rational 'comprehensive' model that assumes that decision-makers have a complete understanding of their goals, the policy problem in question, the options available to resolve it and can rationally weigh up the costs and benefits of each as a comprehensive decision-making procedure. Lindblom (1959: p. 81) rejected this approach. He argued that policymaking does and should proceed by gradual incremental changes through 'successive limited comparisons' which build from the current policy arrangements, step-by-step and by small degrees, without any major shift or step-change in policy direction.

3 The seeming reticence of many in the natural sciences academy to endorse such views at the present time is noteworthy, as positivist viewpoints still predominate, implicitly or otherwise, in ways that assume that science and empirical evidence can unequivocally inform (if not also directly determine) policy decisions. Nonetheless, constructivist frameworks that are amenable to the need for the democratisation of expertise are now widely used by environmental scientists, geographers and social scientists.

4 As I argue here, the point of demarcation between expertise and politics may be difficult to discern due to the epistemic intractability and irreducibility of many contemporary policy problems. Funtowicz and Ravetz' thesis helps to elucidate this epistemic dilemma, but can hardly be blamed for blurring the lines between expertise and politics, which has happened because of the inability of scientists to satisfactorily justify their privileged authority in many public decision-making realms or to provide usable evidence for that purpose.

5 New Public Management refers to efforts since the 1980s to redesign policymaking institutions under the influence of neo-liberal economic reforms that called for a reduction in the size of government (Harvey, 2007). NPM manifested itself in varying ways but most bureaucracies under neo-liberal political influence became more dependent on arms-length government institutions and consultancies for the provision and interpretation of evidence and the development of policy solutions. This process has been referred to as the 'hollowing-out of government' whereby ministerial departments increasingly relinquished direct participation in knowledge production for policymaking (Head, 2008b; Bevir et al., 2003).

6 See Chapter 7 for a more in-depth discussion of the processes of politicisation of science and the scientisation of policy debates.

7 'our attempts to chart the world and its development' (Rip, 2003: p. 423).

8 This summary of Collins and Evans' thesis and the counter-arguments which followed does not do justice to the depth and nuance of ideas involved in their fascinating academic debate. Nor indeed, does this summary do justice to the ideas of philosophers of science, politics and human geography who have made important contributions to understanding how values may legitimately or otherwise influence expert authority concerning the natural world. I hope, nonetheless, that it suffices to provide a brief introduction to the

problems of demarcation relevant in this book and as a basis for understanding the tensions between experts, evidence and policymaking which are explored in later chapters.

9 See for instance, Jasanoff (2003c: p. 394): '[That Collins and Evans try t]o label some aspects of society's responses to uncertainty "political" and some others "scientific" makes little sense when the very contours of what is certain or uncertain in policy domains get established through intense and intimate science-society negotiations'.

Bibliography

Allison, GT 1971, *Essence of Decision: Explaining the Cuban Missile Crisis*, Little, Brown & Co., Boston, MA.

Althaus, C, Bridgman, P and Davis, G 2007, *The Australian Policy Handbook*, 4th edn, Allen & Unwin, Crows Nest, NSW.

Australian Public Service Commission (APSC) 2007, *Tackling Wicked Problems: A Public Policy Perspective*, Australian Government, Canberra.

Bayley, CA 2004, *The Birth of the Modern World, 1740– 1914: Global Connections and Comparisons*, Blackwell, London.

Beck, U 1992, *Risk Society: Towards a New Modernity*, Sage, London.

Bell, S 2004, '"Appropriate' policy knowledge, and institutional and governance implications', *Australian Journal of Public Administration*, vol. 63, no. 1, pp. 22–28.

Bevir, M, Rhodes, RAW and Weller, P 2003, 'Traditions of governance: Interpreting the changing role of the public sector', *Public Administration*, vol. 81, no. 1, pp. 1–17.

Bruner, RD and Steelman, TA 2005, 'Beyond Scientific Management', in Bruner, RD, Steelman, TA, Coe-Juell, L, Cromley, CM, Edwards, CM and Tecker, DW (eds), *Adaptive Governance: Integrating Science, Policy, and Decision Making*, Columbia University Press, New York.

Bucchi, M and Neresini, F 2008, 'Science and Public Participation', in Hackett, EJ, Amsterdamska, O, Lynch, M and Wajcman, J (eds), *The Handbook of Science and Technology Studies*, 3rd edn, The MIT Press, Cambridge, MA.

Cash, D, Clark, W, Alcock, F, Dickson, N, Eckley, N and Jager, J 2002, *Salience, Credibility, Legitimacy and Boundaries: Linking Research, Assessment and Decision Making*, John F. Kennedy School of Government, Harvard University, Faculty Research Working Papers Series, November 2002.

Cash, DW, Clark, WC, Alcock, F, Dickson, NM, Eckley, N, Guston, DH, Jager, J and Mitchell, RB 2003, 'Knowledge systems for sustainable development'. *Proceedings of the National Academy of Sciences*, vol. 100, no. 14, pp. 8086–8091.

Churchman, CW 1956, 'Science and decision making', *Philosophy of Science*, vol. 22, pp. 247-249

Collins, HM and Evans, R 2002, 'The third wave of science studies: Studies of expertise and experience', *Social Studies of Science*, vol. 32, no. 2, pp. 235–296.

Collins, HM and Evans, R 2003, 'King Canute meets the Beach Boys: Responses to the third wave', *Social Studies of Science*, vol. 33, no. 3, pp. 435–452.

Douglas, H 2004, 'The irreducible complexity of objectivity', *Synthese*, vol. 138, no. 3, pp. 453–473.

Douglas, H 2009, *Science, Policy, and the Value-Free Ideal*, University of Pittsburgh Press, Pittsburgh, PA.

Douglas, M and Wildavsky, A 1983, *Risk and Culture*, University of California Press, London.

Dryzek, JS 1993, 'Policy Analysis and Planning: From Science to Argument', in Fischer, F and Forester, J (eds), *The Argumentative Turn in Policy Analysis and Planning*, Duke University Press, Durham, NC.

du Gay, P 2005, *The Values of Bureaucracy*, Oxford University Press, Oxford.

Durant, D 2011, 'Models of democracy in social studies of science', *Social Studies of Science*, vol. 41, no. 5, pp. 691–714.

Durant, D 2016, 'The Undead Linear Model of Expertise', in Heazle, M and Kane, J (eds), *Policy Legitimacy, Science and Political Authority: Knowledge and Action in Liberal Democracies*, Earthscan, Routledge, London.

Dye, TR 2005, *Understanding Public Policy*, 11th edn, Pearson Prentice Hall, Saddle River, NJ.

Etzioni, A 1967, 'Mixed-scanning: A 'third' approach to decision-making', *Public Administration Review*, vol. 27, no. 5, pp. 385–392.

Fischer, F 2009, 'Technical Knowledge in Public Deliberation: Towards a Constructivist Theory of Contributory Expertise', in Fischer, F (ed.), *Democracy and Expertise: Reorienting Policy Inquiry*, Oxford University Press, Oxford.

Funtowicz, SO and Ravetz, JR 1993, 'Science for the post-normal age', *Futures*, vol. 25, no. 7, pp. 739–755.

Gerth, HH and Wright Mills, C (eds) 1970, *From Max Weber: Essays in Sociology*, Routledge & Kegan Paul, London.

Gibbons, M, Limoges, C, Nowotny, H, Schwartzman, S, Scott, P and Trow, M 1994, *The New Production of Knowledge: The Dynamics of Science and Research in Contemporary Societies*, Sage, London.

Gieryn, TF 1983, 'Boundary-work and the demarcation of science from non-science: Strains and interests in professional ideologies of scientists', *American Sociological Review*, vol. 48, no. 6, pp. 781–795.

Gieryn, TF 1995, 'Boundaries of Science', in Jasanoff, S, Markle, GE, Peterson, JC & Pinch, T (eds), *Handbook of Science and Technology Studies*, Sage, Thousand Oaks, CA.

Gross, PR and Levitt, N 1998, *Higher Superstition: The Academic Left and its Quarrels with Science* pbk edn, The Johns Hopkins University Press, London.

Grundmann, R 2006, 'Ozone and climate: Scientific consensus and leadership', *Science, Technology & Human Values*, vol. 31, no. 1, pp. 73–101.

Grundmann, R 2009, 'The role of expertise in governance processes', *Forest Policy and Economics*, vol. 11, nos. 5–6, pp. 398–403.

Guston, DH 2001, 'Boundary organizations in environmental policy and science: An introduction', *Science, Technology, & Human Values*, vol. 26, no. 4, pp. 299–408.

Harvey, D 2007, *A Brief History of Neoliberalism*, Oxford University Press, Oxford.

Head, B 2008a, 'Wicked problems in public policy', *Public Policy*, vol. 3, no. 2, pp. 101–118.

Head, B 2008b, 'Three lenses of evidence-based policy', *Australian Journal of Public Administration*, vol. 67, no. 1, pp. 1–11.

Head, BW 2013, 'Evidence-based policymaking: Speaking truth to power?', *Australian Journal of Public Administration*, vol. 72, no. 4, pp. 397–403.

Heazle, M, Tangney, P, Burton, P, Howes, M, Grant-Smith, D, Reis, K and Bosomworth, K 2013, 'Mainstreaming climate change adaptation: An incremental approach to disaster risk management in Australia', *Environmental Science & Policy*, vol. 33, pp. 162–170.

Hertin, J, Turnpenny, J, Jordan, A, Nilsson, M, Russel, D and Nykvist, B 2009, 'Rationalising the policy mess? Ex ante policy assessment and the utilisation of knowledge in the policy process', *Environment and Planning A*, vol. 41, no. 5, pp. 1185–1200.

Hoppe, R 2005, 'Rethinking the science-policy nexus: From knowledge utilization and science technology studies to types of boundary arrangements', *Poiesis Prax*, vol. 3, no. 3, pp. 199–215.

Howes, M, Tangney, P, Reis, K, Grant-Smith, D, Heazle, M, Bosomworth, K and Burton, P 2015, 'Towards networked governance: Improving interagency communication and collaboration for disaster risk management and climate change adaptation in Australia', *Journal of Environmental Planning and Management*, vol. 58, no. 5, pp. 757–776.

Hulme, M 2009, *Why We Disagree about Climate Change: Understanding Controversy, Inaction and Opportunity*, Cambridge University Press, Cambridge.

Irwin, A 2006, 'The politics of talk: Coming to terms with the 'new' scientific governance'. *Social Studies of Science*, vol. 36, no. 2, pp. 299–320.

Jaeger, CC, Renn, O, Rosa, EA and Webler, T 1998, 'Decision Analysis and Rational Action', in Rayner S and Malone, EL (eds), *Human Choice and Climate Change*, Volume 3: *Tools for Policy Analysis*, Batelle Press, Columbus, OH.

Jasanoff, S 1990, *The Fifth Branch: Science Advisers as Policymakers*, Harvard University Press, Cambridge, MA.

Jasanoff, S 2003a, '(No?) Accounting for expertise', *Science and Public Policy*, vol. 30, no. 3, pp. 157–162.

Jasanoff, S 2003c, 'Breaking the waves in science studies: Comment on H.M. Collins and Robert Evans, 'The Third Wave of Science Studies', *Social Studies of Science*, vol. 33, no. 3, pp. 389–400.

Jasanoff, S (ed) 2004, *States of Knowledge: The Co-Production of Science and Social Order*, Routledge, London.

Jasanoff, S 2005, 'Judgement under Siege: The Three-Body Problem of Expert Legitimacy', in Maasen, S and Weingart, P (eds), *Democratization of Expertise? Exploring Novel Forms of Scientific Advice in Political Decision-Making*, Springer, Dordrecht, the Netherlands.

Jasanoff, S and Wynne, B 1998, 'Science and Decision-Making', in Rayner, S and Malone, EL (eds), *Human Choice and Climate Change*, Volume 1: *The Societal Framework*, Batelle Press, Columbus, OH.Jones, BD and Baumgartner, FR 2012, 'From there to here: Punctuated equilibrium to the general punctuation thesis to a theory of government information processing', *Policy Studies Journal*, vol. 40, no. 1, pp. 1–19.

Juntti, M, Russel, D and Turnpenny, J 2009, 'Evidence, politics and power in public policy for the environment', *Environmental Science and Policy*, vol. 12, no. 3, pp. 207–215.

Keller, AC 2009, *Science in Environmental Policy: The Politics of Objective Advice*, The MIT Press, London.

Kellow, A 2009, *Science and Public Policy: The Virtuous Corruption of Virtual Environmental Science*, Edward Elgar Publishing, Cheltenham, UK.

Kitcher, P 2011. *Science in a Democratic Society*, Prometheus Books, New York.

Kuhn, T 1962, *The Structure of Scientific Revolutions*, The University of Chicago Press, Chicago, IL.

Latour, B and Woolgar, S 1979, *Laboratory Life: The Construction of Scientific Facts*, Princeton University Press, Princeton, NJ.

Leiss, W 2000, 'Between Expertise and Bureaucracy: Risk Management Trapped at the Science-Policy Interface', in Doern, GB and Reed, T (eds), *Risky Business: Canada's Changing Science-Based Policy and Regulatory Regime*, University of Toronto Press, Toronto.

Lindblom, CE 1959, 'The science of "muddling through"', *Public Administration Review*, vol. 19, no. 2, pp. 79–88.

Lyles, MA and Thomas, H 1988, 'Strategic problem formulation: Biases and assumptions embedded in alternative decision-making models', *Journal of Management Studies*, vol. 25, no. 2, pp. 131–145.

Maasen, S and Weingart, P 2005, 'What's New in Scientific Advice to Politics?', in Maasen, S and Weingart, P (eds), *Democratization of Expertise? Exploring Novel Forms of Scientific Advice in Political Decision-Making*, Springer, Dordrecht, the Netherlands.

Marston, G and Watts, R 2003, 'Tampering with the evidence: A critical appraisal of evidence-based policy making', *The Drawing Board: An Australian Review of Public Affairs*, vol. 3, no. 3, pp. 143–163.

Miller, TR and Neff, MW 2013, 'Defacto science policy in the making: How scientists shape science policy and why it matters (or, why STS and STP scholars should socialise)', *Minerva*, vol. 51, pp. 295–315.

Milman, O 2014, 'Scientists give Tony Abbott's record a lukewarm reception', *The Guardian*, Thursday, 30 October 2014, viewed 1 December 2014, www.theguardian.com/australia-news/2014/oct/30/scientists-tony-abbott-record-lukewarm-reception/.

Nilsson, M, Jordan, A, Turnpenny, J, Hertin, J, Nykvist, B and Russel, D 2008, 'The use and non-use of policy appraisal tools in public policymaking: An analysis of three European countries and the European Union', *Policy Science*, vol. 41, no. 4, pp. 335–355.

Nowotny, H 2003, 'Democratising expertise and socially robust knowledge', *Science and Public Policy*, vol. 30, no. 3, pp. 151–156.

Oreskes, N 2004, 'Science and public policy: What's proof got to do with it?', *Environmental Science and Policy*, vol. 7, pp. 369–383.

Oreskes, N, Shrader-Frechette, K and Belitz, K 1994, 'Verification, validation and confirmation of numerical models in the earth sciences', *Science*, vol. 263, no. 5147, pp. 641–646.

Owens, S, Rayner, T and Bina, O 2004, 'New agendas for appraisal: Reflections on theory, practice, and research', *Environment and Planning A*, vol. 36, no. 11, pp. 1943–1959.

Page, E 1985, *Political Authority and Bureaucratic Power: A Comparative Analysis*, Harvester Press, Brighton, UK.

Parsons, DW 1995, *Public Policy: An Introduction to the Theory and Practice of Policy Analysis*, Edward Elgar Publishing, Cheltenham, UK.

Parsons, K (ed) 2003, *The Science Wars: Debating Scientific Knowledge and Technology*, Prometheus Books, New York.

Parsons, W 2004. 'Not just steering but weaving: Relevant knowledge and the craft of building policy capacity and coherence', *Australian Journal of Public Administration*, vol. 63, no. 1, pp. 43–57.

Peters, G 2003, 'Dismantling and Rebuilding the Weberian State', in Hayward, J and Menon, A (eds), *Governing Europe*, Oxford University Press, Oxford.

Pielke Jr, RA 2007, *The Honest Broker: Making Sense of Science in Policy and Politics*, Cambridge University Press, Cambridge.

Price, DK 1965, *The Scientific Estate*, The Belknap Press of Harvard University Press, Cambridge, MA.

Radaelli, CM and Meuwese, ACM 2010, 'Hard questions, hard solutions: Proceduralisation through impact assessment in the EU', *West European Politics*, vol. 33, no. 1, pp. 136–153.

Restivo, S 1995, 'The Theory Landscape in Science Studies: Sociological Traditions', in Jasanoff, S, Markle, GE, Peterson, JC and Pinch, T (eds), *Handbook of Science and Technology Studies*, Sage, Thousand Oaks, CA.

Rittel, HW and Webber, MM 1973, 'Dilemmas in a general theory of planning', *Policy Sciences*, vol. 4, no. 2, pp. 155–169

Rip, A 2003, 'Constructing expertise: In a third wave of science studies?', *Social Studies of Science* vol. 33, no. 3, pp. 419–434.

Roberts, JT 2007, 'Is Logical Empiricism Committed to the Ideal of Value-Free Science?', in Kincaid, H, Dupre, J and Wylie, A (eds), *Value-Free Science? Ideals and Illusions*, Oxford University Press, Oxford.

Sabatier, PA 1988, 'An advocacy coalition framework of policy change and the role of policy-oriented learning therein', *Policy Sciences*, vol. 21, no. 2/3, pp. 129–168.

Sabatier, PA (ed) 1999, *Theories of the Policy Process*, Westview Press, Boulder, CO.

Sarewitz, D 2000, 'Science and Environmental Policy: An Excess of Objectivity', in Frodeman, R (ed.), *Earth Matters: The Earth Sciences, Philosophy, and the Claims of Community*, Prentice Hall, Ann Arbor, MI.

Sarewitz, D 2004, 'How science makes environmental controversies worse', *Environmental Science & Policy*, vol. 7, pp. 385–403.

Schon, DA and Rein, M 1994, *Frame Reflection: Toward the Resolution of Intractable Policy Controversies*, Basic Books, New York.

Sismondo, S 2008, 'Science and Technology Studies and an Engaged Program', in Hackett, EJ, Amsterdamska, O, Lynch, M and Wajcman, J (eds) 2008, *The Handbook of Science and Technology Studies*, 3rd edn, The MIT Press, Cambridge, MA.

Solesbury, W 2001, *Evidence Based Policy: Whence it Came and Where it's Going –Working Paper 1*, ESCRC UK Centre for Evidence Based Policy and Practice, Queen Mary, University of London.

Spruijt, P, Knol, AB, Vasileiadou, E, Devilee, J, Lebret, E and Petersen, AC 2014, 'Roles of scientists as policy advisers on complex issues: A literature review', *Environmental Science & Policy*, vol. 40, pp. 16–25.

Sutton, R 1999, *The Policy Process: An Overview*. Overseas Development Institute, Chameleon Press, London.

Towell, N, Kenny, M and Smith, B 2013, 'Razor taken to CSIRO', *The Sydney Morning Herald*, 8 November 2013, viewed 1 December 2014, www.smh.com.au/federal-politics/political-news/razor-taken-to-csiro-20131107–2x4fu.html/.

Turner, S 2001, 'What is the problem with experts?', *Social Studies of Science*, vol. 31, no. 1, pp. 123–149.

Turnpenny, J, Radaelli, CM, Jordan, A and Jabob, K 2009, 'The policy and politics of policy appraisal: Emerging trends and new directions', *Journal of European Public Policy*, vol. 16, no. 4, pp. 640–653

Weinberg, AM 1972, 'Science and trans-science', *Minerva*, vol. 10, no. 2, pp. 209–222.

Weingart, P 1999, 'Scientific expertise and political accountability: Paradoxes of science in politics', *Science and Public Policy*, vol. 26, no. 3, pp. 151–161.

Weiss, CH 1979, 'The many meanings of research utilization', *Public Administration Review*, vol. 39, no. 5, pp. 426–431.

Wesselink, A and Hoppe, R 2011, 'If post-normal science is the solution, what is the problem?: The politics of activist environmental science', *Science, Technology & Human Values*, vol. 36, no. 3, pp. 389–412.

Wynne, B 2003, 'Seasick on the third wave? Subverting the hegemony of propositionalism: Response to Collins & Evans (2002)', *Social Studies of Science*, vol. 33, no. 3, pp. 401–417.

3 Queensland, Australia and the UK

Comparing the pursuit of climate adaptation in liberal democracies

Whatsoever every man desireth, that for his part he calleth good

Thomas Hobbes

The difficulties of deriving definitive and wholly impartial scientific knowledge for policymaking, as discussed in the previous chapter, suggests that without due care politics can play an active role in evidence development. In this chapter I describe the political, historical and policymaking contexts for the cases examined in this book. Understanding these contexts will allow readers to understand why policy evidence and expertise may have been influenced by prevailing politics to date.

I compare the contrasting cases of Queensland, Australia and England in the UK, both world leaders in climate adaptation policy and practice. In Queensland climate adaptation has, more often than not, been framed in terms of addressing climate exposure, while climate *change* has been a continual victim of partisan politics. Queensland's exposure-based approach has allowed policymakers to effectively ignore climate change science and to exclude consideration of climate change from the evidence base when it has been politically expedient to do so. By contrast England, and the UK more generally, has primarily framed the problem in terms of climate *change* vulnerability and risk. The UK's approach, however, often mandates the use of scientific knowledge in a way that, I argue, allows for tacit approval of political priorities under the guise of impartial expertise.

These cases' contrasting strategies have resulted in a curious outcome arising from their differing uses of scientific expertise for climate risk management. Framing the adaptation problem in terms of society's ongoing exposure to climate has allowed Queensland governments to maintain and enhance the resilience of communities to extreme climate events. Although their short-sighted approach may enhance vulnerability to future climate change over the long term, I argue, resilience-building activities in Queensland appear equivalent to, and have in some respects surpassed, the comparable efforts of the UK to enhance communities' adaptive capacity.

In the UK, by contrast, a vulnerability and risk-based approach under the remit of the Climate Change Act (2008) has placed great emphasis on the development of an evidence base and an institutional framework capable of managing future climate change. The UK's adaptation policies and planning, however, appear to have been considerably less effective at building climate change resilience than their international reputation often gives them credit for. This is due to a number of factors. First, in the UK climate extremes remain a largely abstract risk that local government, infrastructure and community service providers and emergency services have limited experience of managing in practice. Second, the UK's national approach[1] has emphasised the development and use of 'top-down' assessments of risk using climate science that, under a linear-technocratic schema, are prone to being politicised and often lack usefulness and usability for informing climate adaptation. Third, and relatedly, using climate science and derived risk assessments to justify pre-emptive climate adaptation measures in practice often fails government's cost-benefit test in the absence of certainty over future climate change. In the UK, as elsewhere, economic evidence nearly always trumps scientific evidence. The UK's top-down approach to climate risk management, in combination with a bottom-up approach to climate resilience, I argue, has created a *scientised* policy process that places great explicit emphasis on reducing uncertainties in advance of substantive actions, but that actually prioritises a business-case over an evidence base.

The comparative analysis undertaken here relates to the broad structures and functioning of similar three-tier governance structures. As the reader will note, however, both cases' histories in relation to climate, the natural environment and the use of scientific expertise contain an array of apparent contradictions between public and political attitudes that can often be difficult to untangle. This book cannot hope to wholly account for the continually shifting policy landscapes of these jurisdictions, both of which have been subject to significant political upheaval in recent years. My analysis nonetheless provides a snapshot of the types of socio-political, economic and governance issues that are important to consider to understand the interactions between politics, experts and evidence for liberal democratic policymaking.

The comparative cases of Queensland, Australia and the UK[2]

Both Queensland and the UK possess characteristics that make them important case-studies for understanding the adaptation policy problem. To begin, both may be considered leaders in the field of climate adaptation, albeit for very different reasons.

Many of the climate challenges presently faced in Queensland, in terms of the frequency, variability and severity of extreme events, are similar to those expected to be faced in the future in many other parts of the world, including the UK. This is due in part to the prevailing influence of non-annual climatic cycles on the east coast of Australia, providing Queensland with an unusual and often erratic climate regime (McDonald et al., 2010; Flannery, 1994: p. 8). This tropical/sub-tropical region, therefore, provides a view into the types of climate adaptation challenges that many other developed liberal democracies in temperate climates have yet to encounter. Elsewhere, climate adaptation is often considered in the context of relatively benign, temperate conditions, as for example in Western Europe, so that climate adaptation is synonymous with climate *change*, primarily through abstract notions of future risk informed by the scenarios and projections devised by the climate science community. The impacts arising from extreme events under temperate climate regimes, I argue, relate as much to existing societal vulnerabilities associated with poorly designed urban space and over-populated communities, as they do to the severity of the existing climate conditions (EA, 2010a), which pale in comparison to those currently experienced in Queensland (Head et al., 2014; Peters and McEntire, 2014).

As a result, policy assessments of existing and future climate vulnerability in the UK, such as for England, have been undertaken with limited practical exposure of communities and government to the challenges potentially faced as a result of climate change. Queensland, by contrast, has a long history of coping with persistent, variable and alternating climate extremes. These extremes range from chronic drought conditions, such as the Millennium Drought (2001 – 2009) to more acute events such as the Brisbane La Nina floods and Cyclone Yasi in 2010 and 2011 (Heazle et al., 2013). Indeed, as I edit the final draft of this book in December 2016, a large bushfire is consuming Russell Island, just east of Brisbane City. In Queensland, extreme weather conditions are perfectly normal.

Queensland governments' approach to climate adaptation, therefore, has been born through the concurrent policy fields of disaster risk management, natural resource management and urban planning, not as an abstract risk based on projections of future climate, but as an ongoing need to manage these extremes. I argue, therefore, that climate risk management in Queensland addresses a level of immediacy of hazardous events that does not exist in most other developed liberal democracies at this time, but may do in the future as a result of climate change. Queensland's approach to climate adaptation policymaking, therefore, can be usefully understood by framing the problem as one of climate exposure, rather than a problem of climate change vulnerability, as in the UK.

Although England, and the UK more generally, has not had the same experience of managing climate extremes, it has emerged as an international leader in the development of climate change policy in the last decade, both in terms of efforts to reduce greenhouse gases (GHGs) and in adapting to future expected impacts (Preston et al., 2011; Keskitalo et al., 2010; Christoff, 2008; Grubb, 2002). This policy development has occurred despite the fact that the country has a reputation for environmental policy neglect and was long considered by its neighbours as the "dirty man of Europe"[3] (Porritt, 1989). The reasons for this supposed policy neglect have been debated at length and the

Figure 3.1 Queensland and southeast Queensland

answers proposed here also point to why now, in terms of climate change at least, the UK has emerged in the twenty-first century as a policy champion, despite its relative lack of experience in dealing with climate hazards.

Both cases advocate an evidence-based approach to policymaking, at least in theory. Both regions also have a long history of incorporating 'rationalist' ideals into government policy, and indeed the UK was one of the pioneers of the contemporary trend of 'evidence-based' policy (Head, 2008b; Solesbury, 2001), making it an ideal case-study for understanding the development of adaptation evidence. Geographically too, these two regions have interesting similarities. In terms of the specific case-studies used in this book, most of which come from southeast Queensland (SEQ) and southeast England (SEE), these regions are both heavily urbanised and are particularly vulnerable to potential future climate change due to their particular geographies.

SEQ is one of Australia's fastest-growing regions. With an increasingly urbanised, expanding, coastal population stretching from the Sunshine Coast in the north to the Gold Coast in the south, SEQ has been identified as a potential climate change 'vulnerability hotspot' by the IPCC (Hennessy et al., 2007: p. 530). Incorporating a total of 11 local government councils, seven of these have, since the 1970s, begun to merge into a single urban agglomeration that has been coined the "200 km City" (Spearritt, 2009). The 200 km City concentrates approximately 90 per cent of SEQ's 3.4 million inhabitants along the coast in an area comprising about 20 per cent of the region's 22,420 km^2 (Queensland Government, 2016a; ABS, 2012; Spearritt, 2009).

Figure 3.2 The United Kingdom and southeast England

By comparison, SEE is the UK's second most populated of 12 governance regions, with 8.6 million residents in an area covering about 19,000 km², making it approximately twice as populous as SEQ by total area, but equivalent in terms of urban density to SEQ's 200 km City. SEE's extensive urbanisation, low-lying topography and high population density mean that it has also been identified as being particularly vulnerable to climate change, due to the potential for water resource shortages and drought, flooding and sea level rise. Meanwhile, its underlying geology makes its coastline particularly vulnerable to erosion (EA, 2010a; Keskitalo, 2010).

These regions' three-tier system of governance is useful for understanding how policy issues like climate adaptation are governed in practice. Queensland is served by federal, state and local governments that have their origins in structures created in the nineteenth century. Described as the 'Washminster mutation' (Thompson, 1980: p. 32), Australia follows a Westminster style of cabinet government at both federal and state level, while having constitutional oversight by a high court and a senate, reminiscent of the US political system. In the UK, government structures in relation to cabinet, parliament and the House of Lords have been in place for considerably longer, although existing bureaucratic structures in the UK have also only developed since the nineteenth century (Bayley, 2004). UK governance is divided into local, national and (up until their forthcoming departure from the EU) trans-national levels, with the latter contributing a swathe of environmental legislation from the European Commission which the UK have been legally obliged to implement. Alongside their recent referendum decision to depart from the EU, however, the UK has also embraced a model of administration for Northern Ireland, Scotland and Wales which devolves much of national government's democratic representation to sub-national cabinet and parliament. Although these devolved structures have taken on considerable responsibility for climate adaptation-related policy, England does not fall within these devolved jurisdictions.

The partisan politics of climate change

These cases have demonstrated quite different attitudes to the prospect of climate change, to developing climate resilience and adapting to climate, which highlights the importance of historical and contextual factors in the use of evidence for policymaking. To begin, I describe here the development of climate change politics and policy in Australia and the UK, from their ratification of the UN Framework Convention on Climate Change (UNFCCC) to the subsequent development of domestic policy initiatives.

Queensland

Australian international and domestic policy on climate change is characterised by strongly polarised politics and a number of seemingly conflicting socio-cultural, economic and environmental factors. Federal government initially demonstrated some eagerness to address the issue at the signing of the UNFCCC in 1992 (Griffiths et al., 2007; Christoff, 1998). The Hawke-Keating government's optimism and political will for domestic policy responses to climate change may have reflected, to some extent, their lack of understanding of the true character of the problem (see Chapters 4 and 8). Their eagerness perhaps also reflected at the time a wave of public concern for the environment due to a number of related high-profile international issues. Ozone depletion, acid rain, the sinking of Greenpeace's Rainbow Warrior in Auckland, New Zealand and growing concern about global warming had all contributed to enhance public support for environmental policies (Curran, 2011; Christoff, 2005). The Labor government's signature of the UNFCCC in 1992 was an acceptance of the international consensus for a substantial commitment, but domestically was seen by many as inconsequential as it placed few policy demands on the Keating government within political timescales (Crowley, 2007).

However, as Australia was signing the UNFCCC, the tide was already turning. Recession was beginning to bite, and energy and primary industries – which were beginning to realise the importance of lobbying government on climate change – would be ready for subsequent rounds of international negotiation. Five years later, in 1997, at the third UN Conference of the Parties (COP 3), the Howard administration had made a U-turn from the country's previous diplomatic position, refusing to accept legally binding targets, and as Christoff (1998) argues, appearing to exploit the tensions occurring between major Annex 1 (developed) countries. In the dying hours of the conference Australian diplomats managed to achieve a remarkably unambitious target: an 8 per cent increase in GHG emissions above 1990 levels by 2012.

Although opinion polls at the time of the Kyoto summit suggested the Howard government's intransigence toward GHG reduction was out of step with the electorate's views (Bulkeley, 2001; Curran, 2009; Crowley, 2007), a substantial international commitment had clearly become difficult to legitimise (Curran, 2011). It appeared that Australia's negotiators were playing to a domestic corporate lobby that was increasingly concerned about economic imperatives as a result of the recession and at the expense of environmental priorities (Head et al., 2014; Herbohn, 2012; Curran, 2011). In advance of the Kyoto summit, Minister for Environment Senator Robert Hill stated: 'The adoption of a uniform reduction target at the upcoming Kyoto conference would have a devastating impact on Australian industry and its ability to create jobs' (Talberg et al., 2013: p. 7).

An ongoing dependence of Australia's export economy on mining and agriculture has been a pervasive and overriding impediment to international negotiation and domestic climate change policy (Head et al., 2014). Indeed, since signing the UNFCCC it has been argued that Australian governments on both sides of the divide have actively supported the development of traditional energy and other GHG-intensive industries and then used those advancements to achieve dispensations from restrictive emissions targets on the basis of resulting economic hardship (Curran, 2011; Christoff, 2005; Christoff, 1998). Although Australia signed the Kyoto Protocol, the Howard government refused to ratify it and consistently opposed any environmental policy activity that could conceivably diminish the country's economic growth, or that might have endangered an economy that was believed to be particularly vulnerable to external market forces due to its narrow export portfolio (Crowley, 2007; Marris, 2007; Christoff, 2005).

And so, the first federal policy response prompted by signing the UNFCCC, the Ecologically Sustainable Development (ESD) Strategy failed to adequately address the climate change issue. Under the terms of a subsequently derived National Greenhouse Response Strategy (NGRS) set up by the chairs of the ESD process in 1992, industry lobbies had successfully influenced definitions of what constituted 'no regrets' measures for GHG reduction (Bulkeley, 2001) and, as a result, the NGRS was effectively rendered impotent. The NGRS was ultimately replaced by the Howard government's equally ineffective National Greenhouse Strategy (NGS) in 1998 (Talberg et al., 2013).

Howard's reluctance to act decisively on climate change, however, was also influenced by broader concerns over its competitiveness in the Asia-Pacific region, and its trading relationship with the US. Many of its trading partners were exempt from (or in the case of the US, had opted out of) the agreement (Crowley, 2007; Marris, 2007). Ultimately this intransigence may have been their undoing as, ten years later, it has been suggested that Howard's election defeat was due, in part, to being out of touch with the electorate's views on the environment. Certainly, Labor's victory in 2007 was not hindered by their strong policy position on climate change relative to the incumbent Howard government, even though Howard had also finally promised to establish a national emissions trading scheme if re-elected (Beeson and McDonald; 2013; Talberg et al., 2013; Rootes, 2008). It seemed as if climate change was finally about to become a legitimate policy priority.

Within weeks of being elected to office in 2007, Kevin Rudd's government had ratified the Kyoto agreement and established a new ministry of climate change. Rudd subsequently developed a Carbon Pollution Reduction Scheme (CPRS) after a great deal of community consultation. However, it was not to be. The CPRS bill was blocked by the senate in 2009, highlighting the political difficulties alluded to by previous policy initiatives such as the NGRS. As Curran (2011) convincingly argues, Rudd couldn't match his potent political rhetoric – centred on ecological modernisation and climate justice – with the reality of the energy-intensive nature of Australia's export economy. The CPRS attempted to appease both those looking for a strong policy stance on GHG mitigation, and those industry lobbies seeking to maintain international competitiveness. What resulted was a watered-down version of a long-awaited emissions trading scheme (ETS) that fully satisfied no one (Curran, 2011; Christoff, 2010). It granted major concessions to GHG-intensive industries, thus conflicting with Rudd's previously potent rhetoric, while nonetheless considered an active threat to those industries. The resulting loss of public and corporate legitimacy appears to have played a significant part in Rudd's inevitable ousting at the hands of his deputy, Julia Gillard, in 2010 (Head et al., 2014; Curran, 2011).

Following its rejection by senate in August 2009, Rudd's CPRS bill was reintroduced to the House of Representatives in October, following an agreement between Liberal Party opposition leader Malcolm Turnbull and the Rudd government to ensure it passed. However, the agreement was then scuppered by Turnbull's Liberal Party leadership defeat at the hands of Tony Abbott, and the bill was once again rejected in December 2009. The bill was then introduced a third time in early 2010, but following Rudd's removal from Labor leadership, it lapsed after seven months due to the beginning of a new parliament. Almost two years later, Julia Gillard eventually succeeded in passing the Clean Energy Act (2011), which included the establishment of a national ETS, the carbon price for which came into force in 2012. A year later in 2013, however, in an act of defiance against the progressive political and scientific consensus, the newly elected Abbott government immediately began the process of dismantling the Act (Talberg et al., 2013). Conservative politics and industry lobbies, it seems, had once more overcome expert authority and public opinion on climate change.

What is noteworthy in understanding these partisan politics in Australia is that the legitimacy of climate change as a policy priority does not correlate very closely with the public's views. What this difference between policymaking and public opinion points to, perhaps, is the influence of fluctuating economic priorities of government and industry. As demonstrated by the efforts of the Rudd government from 2007 to 2010 and subsequently by the Gillard and Abbott administrations, sufficient political legitimacy for substantive climate change policy has been a continuing struggle. While a majority of the Australian public are clearly concerned by the problem, and are in favour of policy to address it in principle (The Climate Institute, 2015; Reser et al., 2012), the inability of policies such as ETS's to counter the economy's deeply rooted dependence on mining and fossil-fuel intensive industry, mean that the politics and value positions surrounding climate change are extremely difficult to resolve at a most fundamental level. At the time of writing, following a general election in July 2016, public concern about climate change has perhaps never been greater in Australia, yet its absence from electioneering rhetoric and debate from both major parties in advance of the 2016 election was notable (Hudson, 2016). Climate change has become the political hot potato that no one wants to hold.

The rhetoric of ecological modernisation, although initially successful for Kevin Rudd in 2007, has been largely redundant when it comes to reinventing or stabilising the Australian economy (Curran, 2011). In 2013, in his pre-election debate with then Liberal-National Party coalition (LNP) leader Tony Abbott, Rudd once again stressed the importance of diversifying the Australian economy to reduce its reliance on mining and the export of fossil-fuels (Rudd, 2013). Abbott's ultimately successful election campaign in 2013, however, was based on an outright rejection of this idea. Abbott promised to remove the 'carbon tax'[4] introduced by Labor, and to disband the Department of Climate Change when they won the election (Abbott, 2013; Maher, 2011). He made good on these promises, while also scrapping government's climate change advisory body, the Climate Commission (ABC, 2014b; Talberg et al., 2013). In place of the Clean Energy Act (2011), Abbott established a Direct Action Plan – Emissions Reduction Fund to purchase carbon-abatement schemes and encourage 'practical ways of reducing emissions' without directly impacting upon industry (ABC, 2014a). Widely criticised as ineffectual, this policy measure has nonetheless remained in place despite Abbot's ousting from office by Malcolm Turnbull in a leadership coup in 2015. Turnbull's past progressive position on climate change, however, appears, at the time of writing, to be stymied as a result of his tentative hold on the LNP leadership and his need to appease the ultra-right-wing members of his coalition. Climate change policymaking in Australia has stalled until such time as government holds sufficient political capital (or bravery) to advance the cause once more.

Meanwhile, the Newman (LNP) state government of Queensland immediately began dismantling climate change as a policy priority from the beginning of its tenure (2012–2015). It dissolved the Office of Climate Change and withdrew most of the climate change policy which had been developed by the preceding Bligh (Labor) government since 2007. Both the international and domestic politics of climate change appear to have strongly influenced the task of climate adaptation policymaking in Queensland. Much of the aforementioned state policy under the Bligh government related to the explicit task of climate *change* adaptation in a similar manner to the UK's efforts. Yet, in the absence of a bipartisan mandate to mitigate GHG emissions, alongside a concurrent ongoing need to adapt to Queensland's climate extremes, it seems unsurprising that the Newman government removed all commitments to climate *change* adaptation from its policy portfolio, focusing instead on the management of existing climate variability. In turn, this has had an adverse effect on the legitimacy of climate change science for adaptation policy in the state.

Nonetheless, as discussed below, there has been much development within the narrower policy field of climate adaptation to suggest that, although political support for climate *change* may wax and wane depending on partisan political power, there has been and continues to be considerable policy development under the guise of concurrent priorities (e.g., disaster risk and water resource management) that fall under the remit of climate adaptation. While this may preclude much of the explicit hand-wringing about future climate associated with climate *change* adaptation, I argue that the nature of climate change evidence and the types of policy responses available to address it in the face of intractable uncertainties and economic-rationalist imperatives, mean that Queensland's adaptation-related policy outcomes are not significantly less meaningful than those of the UK. Indeed Queensland has a well-seasoned and tested policy framework in terms of the management of existing climate extremes and, despite clear limitations in its focus and design, this policy portfolio has considerably enhanced the region's climate resilience. By comparison, the UK's efforts at climate *change* adaptation remain largely untested and often focused more on reducing uncertainties than on enhancing resilience.

The UK

Climate change policy in the UK has received considerably greater political support than in Australia. This can be explained in part by the country's history of self-sufficiency and resource independence, but also by the influence of neo-liberal ideals upon policymaking which created the right circumstances for climate change to be embraced as a worthy policy priority. This politico-economic backdrop, alongside successive governments' continuing ambitions toward global political leadership, helps to explain why, although in the past the UK has often rejected the available evidence base and dismissed international pressure for environmental stewardship (e.g., through EU policy mandates), where it has been convenient to do so, as in the case of climate change, UK governments have embraced science and been at the forefront of progressive politics and policy (Grubb, 2002).

During negotiations at the Kyoto summit in 1997, the UK played a significant role to effectively bridge the diplomatic gap between the conflicting positions of two of the most central actors, the US and the EU (Damro and Mendes, 2005), to ensure a global agreement could be reached (Grubb, 2002; Oberthur and Ott, 1999). Furthermore, although considered an environmental policy laggard by EU counterparts during the 1970s and 1980s, it was Prime Minister Margaret Thatcher who was one of the first political leaders globally to express concern over the threat of climate change during her Conservative Party conference speech of 1988. This eagerness to embrace climate change as a policy issue was principally as a result of domestic socio-political developments and a desire to capitalise on shifting public opinion about the environment, rather than having anything to do with governments' noble intentions for environmental stewardship. As discussed below, such opportunism has often characterised environmental politics in the UK.

In the late 1980s the Thatcher government had sought to switch its energy portfolio from coal to gas for a combination of economic, political and ideological reasons. It was not out of any concern for the environment, or as a result of the advice from experts in that regard. Indeed, for many years the Thatcher government had steadfastly opposed international pressure to adopt air pollution controls (Hajer, 1995). Thatcher had demonstrated almost no concern for environmental issues until the late 1980s, at which point it became clear that a certain political cache could be developed around the environment to enhance the Tory party's standing in the polls (Gray, 1997). Following election success in 1992, the Tory government, now under John Major, had likewise shown little interest in environmental issues. However, Labour's subsequent election victory under Tony Blair in 1997 marked a turning point.

In advance of the Kyoto summit in 1997, it had become evident that Thatcher's previous manoeuvrings in relation to energy policy would likely be fortuitous for the UK's international standing on climate change, providing an opportunity for global leadership that Blair was keen to capitalise on. The UK was one of the few countries that could report static or declining emissions from a 1990 baseline due to its 'dash for gas' over the preceding decade (Benedick, 2001; Oberthur and Ott, 1999; OECD, 1999). The UK, therefore, provided a strong diplomatic presence in the run up to, and during, the Kyoto summit, with John Prescott, the new environment minister, working to promote an agreement not just between the US and the EU, but also amongst its Commonwealth partners Australia and New Zealand (McCormick, 2002; Oberthur and Ott, 1999).

Even so, promises of action on climate change domestically received little attention during the Blair government's first term in office. McCormick (2002) has suggested that this was mostly due to other ambitious policy commitments in relation to crime, healthcare and education, as well as to the fact that its aspirations toward being the 'greenest UK government ever' (Assinder, 2000) were stymied by the realities of introducing environmental policies that conflicted with existing political and economic priorities – in a rather similar fashion to the failure of Kevin Rudd's climate change rhetoric following election victory in Australia in 2007 (see above). Although there is strong support for climate change policy in the UK and the legitimacy of the associated science, political efforts in this regard have still faltered in terms of finding what McLean (2008) calls 'economically literate' policies. He concludes for this reason that environmental issues, and even climate change, remain an issue of only medium importance in the UK.

Nonetheless, Blair's attention was drawn more toward environmental concerns in his second term in office, particularly in relation to climate change (McCormick, 2002). In 2006 the Labour government commissioned the Stern Review on the economics of climate change mitigation (Stern, 2006) which received global media attention, and in 2008 the Climate Change Act (2008) was passed through parliament with an overwhelming parliamentary majority, thus demonstrating its widespread political support[5] (Moore, 2013a). A pioneering piece of legislation, the likes of which had not been seen anywhere in the world up to that point, the Climate Change Act (2008) was designed to ensure a legislative mandate for the reduction of GHG emissions in a way that could effectively bypass the short-termism of partisan political cycles, committing successive governments to achieving a 80 per cent reduction in emissions by 2050.

Despite significant criticisms in terms of its feasibility and the achievability of its targets (Pielke Jr, 2009; Anderson et al., 2008), the bill is significant because it definitively demonstrated, perhaps for the first time, the magnitude of political will available in the UK to strategically address the problem of climate change. This bipartisan legitimacy appears in stark contrast to similar efforts up to that point in relation to the environment (e.g., for the problem of acid rain – see below), both domestically and at EU level where the UK had traditionally lagged behind and followed the leadership of countries such as Germany and the Netherlands (McCormick, 2002). The UK's leadership on climate change also sits in contrast to the case of Australia where attempts to establish emissions reduction policies have met with persistent political and industry opposition (ABC, 2014b).

Following the passing of the Climate Change Act in 2008, the Tory/Liberal Democrat coalition government under David Cameron came to power in 2010 with the promise of continuing the UK's commitment to tackling the problem. In the preceding years, Cameron had used environmental issues, and climate change in particular, as a device for 'brand decontamination' of the Tories, to move away from its reputation as the 'nasty party' (Carter, 2009). By the time of the election, however, Cameron was reaching the limits of political legitimacy on the environment as the more conservative elements of his party became increasingly annoyed by his embrace of green issues.[6] As a result, he stepped back from environmental messages and concentrated on traditional conservative priorities of crime and healthcare, as well as on issues of public debt and the economic fallout from the global financial crisis since 2008 (Grice, 2009; Carter, 2009). In a similar way, Carter (2009) alludes to a rather ironic twist in the environmental story of the UK.

The Labour governments' development of a renewable energy strategy anticipated a massive increase in the development of both onshore and offshore wind power in order to meet the obligations of the Climate Change Act (2008). Yet this strategy has encountered significant public opposition, particularly from conservative voters, as a result of opposing environmental claims about the defilement of the British landscape. In essence, the difficulties of meeting the Act's GHG targets have been influenced by an ongoing tension between two opposing sets of environmentally conscious political positions: those seeking to preserve the British landscape, versus those seeking to mitigate the global environmental problem of climate change. Indeed, it is noteworthy that the Cameron government's first environment secretary Caroline Spelman had previously called for a moratorium on the construction of wind farms in Scotland in an attempt to appease conservative voters (Toynbee, 2007).

Vehement local opposition to renewable energy infrastructure often encountered in the UK is further testament to the difficulties faced in matching climate change rhetoric with effective policy action, even where there is credible evidence to support it. Despite these difficulties, however, cross-party consensus concerning the need to tackle climate change has remained[7] (Carrington, 2015). The Climate Change Act (2008) obliges government to pursue an evidence-based approach to both the regulation of GHGs and the management of risks associated with climate change. Moreover, and as I discuss below, it is noteworthy that the evidence-based requirements of the Act have further legitimised state-funding for the development and use of ambitious climate change science outputs.

It seems, therefore, that while climate change has consistently been at odds with Australia's political and economic objectives, it has often conveniently aligned with the objectives of successive UK governments. As we shall see, however, the socio-cultural and economic structures underpinning these cases can tell us much about the politics of climate adaptation and the strength of the legitimacy of climate change expertise for this purpose. In order to put the politics of climate change in perspective, therefore, I now provide a historical account of public and political attitudes toward the natural environment as a means to more effectively understand the adaptation problem for these two case-study regions.

Historical, political and socio-economic influences on adaptation policy

Queensland

The Australian approach to environmental management and policy has been characterised by entrenched and often conflicting forces relating to socio-cultural norms, identity, economy and the apparent rationality of the evidence-based mandate. Across three levels of government, there has rarely been a consistent approach to environmental management.[8] In part this is because, although individual states are dependent on federal government for a share of national tax revenue and certain sets of guidance, responsibility for the provision of services and infrastructure largely falls to state and local government. Similar to environmental management between EU member states in the case of the UK, the six states and two main territories of the Australian federation, more often than not, have pursued independent strategies in a way that has been problematic when faced with issues that transcend these jurisdictions (Mercer et al., 2007; Walker, 2002).

This difficulty is compounded by a general lack of integration and coordination and ineffective devolution of authority between governance structures that has undermined relations between federal, state and local levels (Howes, 2008; Christoff, 2005). Cross-level governance has been characterised by power struggles; conflict and mistrust as a result of overlapping jurisdictions; a lack of constitutional recognition for local government, and accusations of cost-shifting[9] (Howes and Dedekorkut-Howes, 2010). This lack of policy integration across scales can exacerbate ongoing difficulties associated with achieving political consensus for climate change policies (Howes, 2008; Toyne, 1994). In these circumstances, adaptation policy problems and therefore policy evidence development may look very different from the perspectives of differing governance levels.

Similar to the UK, Queensland's governance often relies on linear, rationalist approaches to disaster risk management, urban planning policy and climate adaptation (Heazle et al., 2013; Howes et al., 2012). This approach is demonstrated, for instance, in government guidance for assessing natural disaster risks, the operation of its natural resource and disaster risk mitigation assets (Heazle et al., 2013), and its use of Annual Exceedance Probabilities as flood risk management thresholds for urban planning (Tangney, 2015; QFCI, 2012). Alongside governments' contemporary adherence to this rationalist model, and its explicit pursuit of evidence-based policy (Productivity Commission, 2010), however, Queensland's diffuse rural settlement patterns, socio-cultural history and prevailing political landscape suggest rather conservative attitudes toward the natural environment. In particular, Queenslanders have a history of mistrust for the role of experts in managing the environment, and therefore, the extent to which experts do or should have authority over policymaking.

As discussed above, although opinion polls suggest that the Australian public are concerned about anthropogenic climate change and concede that it requires policy action (The Climate Institute, 2015), they (or their political representatives) are often unwilling to accept policies that may endanger their economic prosperity in any way. Public opinion was certainly swayed by the economic case for abandoning the Clean Energy Act when Tony Abbott's coalition gained power in 2013. Moreover, the public and the media have often eschewed, or at least been apathetic toward the advice of the scientific community in relation to environmental problems associated with water resource use and environmentally sound land management practices (see for example, Mercer et al., 2007; Devine, 2002; Dovers, 2000: p. 9). These public attitudes may have something to do with a conflict between the socio-cultural and economic foundations of Australian society on the one hand, and the changing nature of expert advice for environmental problems, on the other.

Contemporary expert knowledge concerning the management of Australia's natural environment has moved beyond anthropocentric notions about our ability to control environmental variables and to dominate ecosystems, toward an increasing acceptance of the need to work within the thresholds of what any given social-ecological system can sustain (Curran, 2011; Howes et al., 2010; Bell, 2006). The Australian public, however, has often expected that state government should support and protect the public from environmental risks through the provision of infrastructure and engineering technologies (along with, presumably, the expertise associated with delivering these technologies) to reduce the public's exposure and/or vulnerability to natural resource shortages and climate extremes (Heazle et al., 2013; Mercer et al., 2007). It would appear, therefore, that the public is on-board with the evidence-based mandate and reliant on technical expertise for climate risk management in terms of infrastructure and engineering solutions, yet sceptical and reluctant to heed expert advice when it suggests the need for restraint, economic risk or the existence of limits in relation to how society can and should interact with the natural environment.

This apparent contradiction in public attitudes to expert authority, I suggest, relates partly to Queensland's traditions of political authority. Australia has had a long tradition of interventionist government and reliance on an expansive public sector (Mercer et al., 2007; Wanna and Weller, 2003). This tradition originated from the difficulties of colonising such a large land mass with a diversity of hostile climates and landscapes (Evans, 2007: p. 28; Johnston, 1988). These circumstances prompted a sort of 'statist developmentalism' (Crowley and Walker, 2012; Walker, 2002: p. 255); the private enterprise of European settlers relied heavily on state support for the social and economic infrastructures necessary to establish viable yet disparate communities across a vast country. Mercer et al. (2007) argue that statist developmentalism has driven natural resource use and climate risk management since the very beginnings of European settlement of the continent in 1788, largely to the exclusion of broader considerations about the contribution of ecosystem services to sustaining socio-economic prosperity.

Queensland's reliance on infrastructure and engineering for environmental management, and its seemingly contradictory position on the legitimacy of expertise, however, must be considered alongside another key characteristic of Australia's cultural identity; one that sits in partial contradiction to statist governance traditions: the colonial culture of 'the stoic battler, surviving through hard work and persistence in a battle against the environment' (Bell, 2006:

p. 562). Throughout much of its history, colonial settlers in Australia sought to overcome the extremes of climate and landscape, by applying European attitudes and practices to agriculture, land management and natural resource use (Fitzgerald, 1984: p. 71). Nature was something to be overcome, something that could be 'made over' (Walker, 2002: p. 252). Native flora and fauna were considered inferior, an impediment to development and growth, to be replaced with European species wherever possible (Mercer et al., 2007). There was a corresponding unwillingness to recognise ecological limits, the character of prevailing climates, or any possible restrictions on socio-economic growth.

Australia's statist-developmentalist tradition in relation to disaster risk management and climate adaptation prevails to this day (Heazle et al., 2013; Mercer et al., 2007), notwithstanding more recent trends toward neo-liberal[10] market-based solutions for environmental management (Pusey, 1991; Garnaut, 2008). And yet, despite this dependence upon infrastructure and engineering to shelter the public from climate extremes, the culture of the 'stoic battler' may have maintained one vital aspect of climate resilience in Queensland that has acted as a powerful counter-current to statist traditions of infrastructure provision and disaster 'relief'. The resilience of Queensland communities to climate extremes has often been noted (see for example, QFCI, 2011, 2012; Howes et al., 2013; Peters and McEntire, 2014). This resilience becomes apparent not only in the provisions for pre-emptive disaster risk and emergency management set out by multi-level government policies (see below), and as repeatedly executed by local government and emergency services, but has also been apparent in the 'social capital' or community adaptive capacity repeatedly demonstrated by Queenslanders in the face of successive extreme events in recent decades (QFCI, 2011, 2012; Howes et al., 2013).

As an indication of this, in 2010 Queensland's Department of Community Safety boasted an army of some 85,000 emergency service volunteers across all agencies (Peters and McEntire, 2014: p. 27) and in the immediate aftermath of the 2011 floods, in Brisbane alone some 20,000 volunteers showed up to coordination points across the city (QFCI, 2011: p. 27). Moreover, it seems that Queensland's policy players increasingly realise the importance of highlighting and enhancing this community resilience and public awareness of climate extremes, perhaps due to economic-rationalist motives, but perhaps also in tacit acceptance of the limitations of statist traditions of infrastructure provision, and in a way that increasingly incorporates and seems to capitalise upon the ethos of the 'stoic battler' (Howes et al., 2013, 2015).

The culture of embattlement over nature, the public's mistrust of contemporary scientific advice for social-ecological management, its dependence on infrastructure and engineering provisions, and its concurrent but contrasting ability to maintain strong local governance and social capital in the face of extreme events, suggests complex tensions between the public, politicians and experts when considering the development of climate adaptation policymaking in Queensland. These tensions have been maintained and enhanced by two further characteristics of this case-study.

The first is the nature of the region's climate itself which is considerably more unpredictable and prone to extremes than those which the first colonists, primarily from Great Britain and Ireland, were used to (Evans, 2007; Johnston, 1988). Conditions of drought, flooding, storms and bushfires are significantly more prevalent in Queensland than in Western Europe, and are subject to greater unpredictability as a result of the influences of the El Nino/Southern Oscillation, alongside longer term episodes of Inter-decadal Pacific Oscillation. While the concept of embattlement over nature is understandable in the context of the region's landscape and colonial history, it takes on considerably greater significance in this region than in many other parts of the world, given that annual variation is not the most significant contributor of weather variability, and given the scale of communities' subsequent exposure to contrasting extremes relative to similarly developed economies in other parts of the world (McDonald et al., 2010; Flannery, 1994). Perhaps Queensland's relative nonchalance toward climate *change*, alongside its climate-resilient social capital, therefore, can be understood in this context.

The second characteristic that contributes to tensions between experts, politics and the public is the historical reliance of the region's export economy on agriculture and mining. This dependence has promoted a continuation of government-supported exploitation of natural resources and has made successive governments mistrustful of international agreements or expert advice that might endanger this economy through, for example, concessions to environmentally sustainable practices (Mercer et al., 2007; Walker, 2002). These characteristics underline the difficulties being faced in this region in achieving political legitimacy for climate change, for international agreements such as the Kyoto Protocol, or for domestic policy responses that might compromise key industry sectors (Curran, 2011).

The UK

The UK has a rather more convoluted history of reliance on government and state services, and yet has encountered similar difficulties and contradictory influences as those in Queensland when attempting to integrate policy between three levels of governance. Richards and Smith (2002) and Harling (2001) recount how, during the nineteenth century, the UK had adopted a laissez-faire economic liberalism whereby government lacked functional ministerial departments and policies were, as often as not, developed either by ministers themselves or by 'policy boards', the latter being increasingly criticised for their lack of democratic legitimacy. Up until the 1860s, government's role was limited to maintaining individual freedoms, private property rights and the market economy, and ensuring protection from external threats. It was as a result of the country's rapid and increasing industrialisation during the nineteenth century that an expansion of the state was ultimately deemed necessary. Industrialisation brought problems of health, social welfare and urban planning that necessitated increased state intervention.

Such intervention became even more pronounced as a result of the first and second world wars, due to the need for careful control of the production and utilisation of resources during these periods. Thus developed an ethos in Britain of national self-sufficiency and resource independence (Sheail, 2002; McCormick, 2002).

After World War II the Labour government of the time embraced Keynesian economic policy that nationalised many infrastructure and key state service providers. However, in response to an economic downturn in the early 1970s that culminated in economic 'stagflation'[11] under the Heath government, in 1979 Margaret Thatcher embraced an ethos of neo-liberal economics that sought to reduce the size and influence of government, to privatise national infrastructure and service providers, and to deregulate industry in order to facilitate the 'free hand' of the market (Harvey, 2007; Harling, 2001). By contrast, Queensland's embrace of neo-liberal ideology during the same period had not stymied its statist-developmentalist norms of rationalist climate risk management policy and that had fomented an expectation on government to eliminate climate risks (Heazle et al., 2013). In the UK, however, Thatcher oversaw the shrinking of government under the auspices of a New Public Management (NPM) regime (see Chapter 2) that sought to farm out many public services to private enterprise (Lane, 2000). This included some of those state services most relevant to climate adaptation, such as the privatisation of water companies in 1989 (OFWAT, 2014b).

The extent of neo-liberal politico-economic influence has remained high since Thatcher's first moves in this direction, despite an expansion of non-departmental government agencies under the Blair government in the 1990s and 2000s. Since 2010, Labour's expansion of government was scaled back by David Cameron's coalition government (Taylor-Gooby, 2012), mirroring neo-liberal austerity policies across Europe during this time (Dunn, 2014). Neo-liberal socio-economic policies have played a significant part in the legitimacy granted to environmental policy issues in the UK during the latter part of the twentieth century and has also influenced contemporary legitimacy for evidence-based adaptation policy at local government level.

The UK's approach to environmental policy, planning and regulation as well as its more recent enthusiasm for climate change policy can thus be explained in the context of its history of rapid and extensive industrialisation and urbanisation in the nineteenth and twentieth centuries; the influence of two world wars that fomented a socio-political culture that placed significant value on resource independence and self-sufficiency, alongside the contemporary prevalence of neo-liberalism (Richards and Smith, 2002; Sheail, 2002; McCormick, 2002; Hajer, 1995). But, just as in Queensland, 'the British approach to environmental policy is notable for its contradictions' (McCormick, 2002: p. 121). Despite its poor international reputation on the environment, the UK has one of the oldest and most established bodies of environmental law in the world, along with one of the world's oldest and largest professional environmental lobbies, and a public with very high levels of environmental awareness and interest in the amenity value of nature (McCormick, 2002; Sheail, 2002; Garner, 2000; Gray, 1997).

Sheail (2002: p. 2) explains how Britons have long been keen to tell the story of their country's landscapes and that this knowledge and understanding strengthened public opinion in favour of the need to conserve and protect the natural environment in the face of rapid urbanisation and industrialisation. As a result, and in contrast to the comparative case of Queensland, natural limits to social and economic growth have long been recognised in the UK. The industrial revolution had caused severe degradation of the British landscape along with a huge increase in the proportion of the expanding population living in urban locations. Both rural and urban locations were reshaped both in terms of their physical and social make-up. As a result, town and country planning provisions developed incrementally from local to regional and eventually to national policy provisions from the 1930s onwards (Rydin, 2003). Likewise, woodland areas were set aside not just to increase the provision of timber resources during the wars, but also for the benefit of rapidly expanding environmental and natural heritage advocacy groups that had sprung up all over the country since the 1850s, and which continue to expand their membership to this day (Evans, 2003; Sheail, 2002). As Sheail (2002: p. 103) argues: 'The spread of the factory system, and subsequent growth of huge towns, had strengthened rather than weakened, the love of things rural.'

In the context of the public's love affair with the natural environment, alongside Thatcher's 'dash for gas' in the 1980s and related concerns for resource self-sufficiency, it seems unsurprising that UK policy players have granted considerable explicit legitimacy to the issue of climate change, and by association, legitimacy for the consensus evidence base for anthropogenic climate forcing. And yet, despite the value placed by the public on the natural environment, existing political-historic analyses suggest that UK governments, much like their Australian counterparts, have rarely placed environmental issues at the top of the political agenda, relative to, say, the economy, healthcare or crime. More commonly, UK policymakers have taken a reactive rather than proactive approach and have often failed to consider environmental issues in a strategic way (Jordan, 2002b).

In this regard, climate change policymaking in the UK cannot be fully understood without an appreciation of its tense relationship with the European Union (EU) (Lowe and Ward, 1998). And indeed, its recent decision to split with the EU could have significant impacts on UK environmental policy provisions given that much of it has either originated from, or been subsumed within EU directives. The UK joined the EU (what was then the European Economic Community (EEC)) in 1973 at a time when the few environmental policies that existed at EU level were adopted by member states in an ad hoc fashion, in accordance with their individual economic and political circumstances. However, following recession in the 1970s European integration stalled, particularly in relation to economic policy. The European Commission

(EC) therefore sought to inch forward integration through the pursuit of technical minutiae and 'low politics' associated with environmental policy. It was on environmental issues where the EC could effect integration in the relative absence of scrutiny from member states and interest groups (Jordan, 2002a). Pressure on the UK to implement EU environmental policy gradually increased during the 1970s and 1980s, so that by the beginning of the 1990s, the UK's department of environment had become so concerned with the prescriptions being sent from Europe that it had established a special coordinating division for EU legislative work (Garner, 2000; Sharp, 1998).

Nonetheless, the UK often vociferously opposed EU pressure in relation to environmental policy that sought to establish a common set of European ideals (McCormick, 2002; Garner, 2000; Lowe and Ward, 1998). Part of this tension may have been as a result of the UK's legacy as a former global power, whereby acceptance of European integration would have come with a concurrent recognition of its status as merely an (upper-)middle-ranking EU member state (Lowe and Ward, 1998). Furthermore, as an island the UK has often failed to understand the impetus of its neighbours toward integration and it has been unsympathetic to economic initiatives that sought to maximise benefit across the land mass of continental Europe. The perceived benefits of, for instance, the EU's Common Agricultural Policy have rarely been as clear-cut to UK diplomats as to their European neighbours (Rydin, 2003: p. 123; Lowe and Ward, 1998). The UK's atypical adversarial political system, its widely recognised bureaucratic efficiency and lacklustre industry regulation, have meant that it was ill-prepared for the requirements of coalition and consensus building that characterise EU policymaking, or the prescriptive nature of the EU's regulatory control (Jordan, 2001).

Garner (2000) and Jordan (2002b, 2001) suggest the UK's difficulty in Europe was primarily as a result of this conflict in policy tradition. The UK approach to industry regulation avoided prescriptive policies and often involved voluntary, pragmatic agreements which were often criticised for failing to effectively regulate polluters. McCormick (2002: p. 125) calls this the UK's 'directionless consensus' between industry and the state.[12] By contrast, the European approach has involved relatively rigid prescriptions and standards to be met by all member states through an evidence-based approach which has been a source of ongoing conflict amongst UK policymakers (Jordan, 2002b).

The oft-quoted example in this regard is the issue of acid rain, which caused significant tension between the UK and its northern European neighbours during the 1970s and 1980s (Lean, 1998). Although a different order of environmental problem (in terms of geographic and temporal scale, as well as their levels of comparative complexity and uncertainty), the UK's management of the acid rain problem is nonetheless a useful example for understanding governments' subsequent approach to climate change since these two issues bear a number of important similarities. Both are cross-border issues requiring international pollution control, both depend on the UK's energy portfolio and its regulation for their resolution, and both have been instrumental in shaping Britain's international relations and reputation on the environment. Furthermore, the example of acid rain explains some of the difficulties that the UK has encountered with cross-level, trans-national governance.

Since first brought to the attention of the EU community (then the EEC) in 1972, it became increasingly clear that sulphate emissions from coal-powered electricity generation in the UK were causing significant acid deposition in countries such as Sweden and Germany (McCormick, 2002; Garner, 2000; Gray, 1997; Hajer, 1995). Yet, despite the considerable evidence to indicate the origins of this pollution, the UK refused to introduce controls to limit sulphate emissions until the late 1980s, at which point more enticing economic and ideological incentives for restrictions on the coal industry had finally swayed government. These new incentives made political space for hitherto unacceptable regulation of an energy source that was, in any case, already on its way out (McCormick, 2002; Hajer, 1995).

The UK's reticence in managing the acid rain problem related principally to two political-economic and cultural issues. First, the country's post-imperial political culture of self-sufficiency in relation to natural resources and agriculture, which had developed as a result of its involvement in two world wars, and had been exacerbated by the global oil supply crises in 1973 and 1979 when dependence on OPEC[13] oil producers had become increasingly and worryingly apparent. Such concerns ensured the UK's obstinate reliance on its abundant coal resources during the 1970s (alongside a developing nuclear energy portfolio) and its rejection of calls for environmental controls that could endanger the economic viability of the energy industry. Second, during the 1980s the Thatcher government, seeking to advance neo-liberal ideals, was even more reluctant to introduce regulatory controls on the energy industry that were seen as contrary to reforms that sought to roll back the state (Sheail, 2002; Hajer, 1995).

These political imperatives appear to have held significantly greater political capital than the need to appease its European neighbours or to address the considerable evidence base suggesting the need for air pollution control (Hajer, 1995). However, by the 1980s the political landscape in the UK was changing, and thus, so too would its relationship with Europe. As part of her neo-liberal agenda, Thatcher sought to rationalise government, to privatise industry and to break the power of industrial unions. A significant step in achieving these goals was to reject the demands of one of the strongest unions in the country, the National Union of Mineworkers, and to close a number of coalmines in northern England with the loss of 20,000 jobs. This resulted in considerable civil unrest and a significant reduction in union membership, but also a need to move toward alternative energy resources (BBC, 2004; Harling, 2001). Thus, at the same time as fossil fuel imports were rapidly increasing due to these mine closures, Thatcher privatised the gas industry and made a major policy shift toward using abundant gas resources in the North Sea (National Gas Museum, 2013). Natural gas presented a cleaner and more economically enticing alternative to coal (Gray, 1997).

Furthermore, much as in Australia during the same period, the UK public were becoming increasingly concerned about environmental issues as a result of a range of high-profile environmental disasters (McCormick, 2002; Garner, 2000; Gray, 1997). This shift in public attitudes meant that there were increasing calls for a move away from coal-fired electricity generation. As the public and political mood was shifting it is little surprise in hindsight that, toward the late 1980s Thatcher became interested in environmental issues. In 1988 Thatcher proclaimed the Tories to be a party of conservationists, as the natural guardians of the environment, and made specific mention of climate change during her Conservative Party conference speech that year. Thus, in 1988 Thatcher also finally ceded to the demands of EU neighbours by agreeing to implement the EC's air quality standards to ensure a reduction in sulphate emissions from the UK's remaining suite of coal-fired power stations, as well as control a range of other dangerous atmospheric pollutants (Garner, 2000; Gray, 1997; Hajer, 1995). Similar to the comparative case of Queensland, this example neatly demonstrates how socio-economic priorities and established or prevailing norms and priorities can supersede the legitimacy of expert evidence and environmental legislation that conflicted with those priorities.

Lowe and Ward (1998: p. XV) nonetheless argue that the EU has had 'a profound and pervasive influence on [British] environmental policy'. Although its influence has not gone so far as to prompt a root and branch reform of the UK's environmental governance structures – which have often been criticised for their inability to strategically manage environmental issues – it would appear that the EU has been instrumental in transforming the UK from an environmental laggard into a relative leader (Gray, 1997; Garner, 2000; McCormick, 2002; Rydin, 2003). However, since the early 1990s, there has been a significant reduction in EU environmental activity, with the Water Framework Directive (2000) the last truly prescriptive piece of legislation to emerge.[14] The biggest casualty of this reduction in legislative activity at EU level during the 1990s may have been efforts to reduce GHGs. In particular the EU failed to implement a carbon tax during this period as part of a European strategy to meet its commitments under the Kyoto Protocol (Garner, 2000). However, since that time there has been considerable progress in this regard, in no small part as a result of the influence from the UK (Grubb 2002; Oberthur and Ott, 1999). As described above, the UK's progressive stance on climate change appears intimately related to socio-cultural and economic norms that value self-sufficiency and international recognition in geo-political affairs.

The UK's tense relationship with the EU in relation to the issue of acid rain, and conversely its progressive stance on climate change, suggest that the legitimacy of both policy and associated evidence is ultimately dependent on the prevailing values and priorities of government. Those priorities for energy independence and international leadership explain why the evidence and policy from the EU was in one case rejected, and in the other embraced and promoted, as a result of their respective congruence with those prevailing values and politics. The historical, political and socio-economic forces described here provide context for understanding the similarities and differences between Queensland and the UK in their development and use of policy evidence, to be explained in later chapters. These case-studies have been subject to similar challenges of cross-level governance yet their historical and socio-economic backgrounds have led to very different sets of political imperatives and policy framings. These differences suggest good reasons for why Queensland and the UK have approached the climate adaptation problem in such different ways and have such divergent levels of legitimacy for climate change science.

The development of climate adaptation policy

Queensland

The issue of climate change, just as in other parts of Australia, has been politically contentious in Queensland. Around the same time that federal government under Tony Abbott was dismantling climate change policy in 2013, at state level the Office of Climate Change established by the Bligh (Labor) government in 2007, was disbanded by the Newman (LNP) government, alongside a review of all state and regional climate change and related policies (Norman, 2012). At local level, climate change policy and planning has been under similar pressure (Killoran, 2012) and state government removed all legal requirement for local councils to incorporate climate change-induced sea level rise into their planning schemes. Indeed, under the Newman administration (2012–2015), Queensland's Minister for Infrastructure and Planning Jeff Seeney issued a formal order to Moreton Bay council to remove all reference to anthropogenic sea level rise from their regional plan (ABC, 2014c), purportedly describing climate change as a 'semi-religious belief' (ABC, 2015).

As discussed above, this dwindling legitimacy may be explained by the ongoing conflict between the conclusions of climate science and prevailing values and dependence of Queensland's economy on fossil-fuel intensive industry and mining, as well as on ongoing trends of coastal urbanisation in the region. Yet, more than ever before, Queensland requires robust climate adaptation policy, as evidenced by a continuing string of extreme climate events in the region in recent years (Howes et al., 2015). Despite these political difficulties, therefore, there have been a surprisingly large number of policy and planning initiatives in the past decade at all governance levels to address climate adaptation-related issues in Queensland.

It seems that although climate *change* adaptation policy has been a continual victim of partisan politics, climate adaptation has nonetheless proceeded apace. Policy development has often occurred under the guise of disaster risk and emergency management and urban planning provisions which don't carry the same political divisiveness as climate change and climate science. Queensland's policies focus on enhancing the resilience of communities to existing climate extremes rather than seeking to legitimise policy using climate change science.

This approach, I argue, has nonetheless yielded considerable benefits for the region's adaptive capacity in the face of climate change.

Adaptation goals that are achieved through concurrent policy priorities have ensured a range of institutional and community resilience-building provisions that under a different regime (such as in the UK), might have been subsumed under climate *change* adaptation policies. The suite of adaptation policy initiatives committed to by state and local governments, irrespective of their political persuasion, underline the ongoing necessity for this policy focus. By contrast, and mirroring a similar trend in policy efforts in the UK, bipartisan commitments toward climate *change* have only been sustained in relation to the development of more and better research and the reduction of uncertainties (Dedekorkut-Howes et al., 2010). As I argue for the UK case below, however, the tactic of committing to uncertainty reduction, though it may sound progressive, often has limited utility for adaptation policymaking and implementation in practice.[15]

The development of explicit climate change adaptation policy in Queensland

The federal government's Department of Climate Change, which existed between 2008 and 2013, prior to the election of the Abbott government, set out a policy paper in 2010, positioning itself as the facilitator of climate change adaptation policy but assigning responsibility for adaptation itself to state and local governments (DCC, 2010). The Council of Australian Governments[16] (COAG, 2007) meanwhile set out a strategy to reduce uncertainties through coordinated research provision and dissemination, with the explicit aim of reducing the vulnerability of key sectors to climate hazards. This strategy has been implemented by the National Climate Change Adaptation Research Facility (NCCARF) since 2008. The Abbott government, although removing climate change as a named priority from ministerial departments (Talberg et al., 2013) and repealing the Clean Energy Act (2011) (ABC, 2014b), nonetheless maintained research and support initiatives through its Climate Change Adaptation Program (Australian Government, 2014a). Initiatives to develop climate change and adaptation science research, however, belie a lack of use of this science for adaptation policy at state and local levels.

Queensland's Office of Climate Change was established by the Bligh (Labor) government in 2007 within what was then the Department of Environment and Resource Management. In 2009 the state government published a climate change strategy (Queensland Government, 2009b) which included 62 adaptation actions relating to water planning and services, agriculture, human settlements, natural environment, emergency services, human health and tourism, business and industry. Most of these actions focused on research initiatives and 'capacity building' (Dedekorkut et al., 2010). This strategy was complemented by a Coastal Plan in 2012 (Queensland Government, 2012a), designed to inform regional and local government planning schemes and development applications across Queensland, in order to prevent development in areas at risk of coastal hazards such as erosion, storm surge or sea level rise.

Following state election success for Campbell Newman's LNP in March 2012, however, most of these climate change policies were scrapped. The Coastal Plan was reissued in a form that referred to climate variability and sea level rise, but made no mention of climate change or global warming. Clearly, the Newman government had downgraded climate *change* adaptation as a policy priority. In 2016, and at the time of writing, the Paluszczuk (Labor) government, predictably, has re-embraced the climate change agenda for state government, and is developing a new Queensland Climate Adaptation Strategy (Queensland Government, 2016b). The absence of bipartisan consensus on climate change, however, makes adaptation a problematic policy focus over the long term, at least until many of the broader issues applicable to its resolution – such as the economy's dependence on mining and fossil-fuel intensive industry – are resolved.

For individual local governments, there are also a number of guidance documents produced at national and state level to direct planning schemes. Bajracharya et al. (2011) have even suggested that the design and focus of adaptation actions at local scale has, in the past, gone some way to bridging the policy gap between national, state and local levels of government. There have been, for instance, individual strategies and plans prepared by Brisbane (BCC, 2007) and Gold Coast (GCCC, 2009) city councils. However, the status of local plans, in light of successive shifts in state and local governments' political priorities, appears precarious at best.

Exposure breeds resilience: Concurrent policy priorities for Queensland

Although the shifting political landscape described for Queensland highlights the precarious position that climate change holds as a policy issue, the nature of the task at hand means that concurrent policy and planning priorities in Queensland often address climate change adaptation, even where it is not mentioned or intended. In this regard, Queensland is in a relatively unusual situation. Given the existing frequency, variability and severity of climate extremes in the region, existing infrastructure provision, urban planning and emergency management organisations and governance mechanisms already deal with a range of climate-related threats with sufficient frequency that local and state government and public services have significant first-hand experience of managing them.

Climate change adaptation is ultimately about enhancing the resilience of communities and the social-ecological systems of which they are a part, to ensure they can withstand the impacts of future climate change[17] (Nelson et al., 2007). Many of the policy and planning activities undertaken at all levels of Queensland's governance, however, as well as within communities themselves that are being undertaken to ensure sustainable development and community safety, appear to enhance the resilience of the region to both existing and future climates. In particular, disaster risk management in Queensland has sought to increasingly enhance community preparedness for, and to reduce communities' vulnerability to, climate variability and extremes. This is achieved through community engagement and education programmes, early warning systems and emergency planning and coordinated disaster response (AIDR, 2016; Queensland Government, 2016c,d; Volunteering Queensland, 2015), as well as by addressing the ongoing pressures of water resource shortages and population growth (QFCI, 2011,2012).

Governance of climate hazards has also been the focus of significant public scrutiny, as well as academic research across Australia in the last ten years (QFCI, 2012; VBRC, 2010; Ross and Dovers, 2008). Because Queensland has been subject to a series of climate-related disasters that have tested existing government infrastructure and institutions to their limit, the difficulties of relying on infrastructure assets and governance programmes that can eliminate climate exposure entirely, have become increasingly apparent (Heazle et al., 2013; Queensland Government, 2014b). An extended period of drought between 2001 and 2009, the La Nina-related flooding events of 2010/2011, the subsequent impacts from Cyclone Yasi shortly afterwards, followed by further flooding events as a result of Cyclone Oswald in 2013 have provided ample demonstration of the destructive force of existing climate variability in Queensland, and in the southeast Queensland region in particular.

In preparation and response to such incidents, Emergency Management Australia (Australian Government, 2016), the principal federal organisation responsible for advising on and overseeing disaster risk management, promotes an 'all agencies', 'all hazards' approach which means that individual hazards should not be managed in isolation from others, or solely by individual government bodies. Through its National Strategy for Disaster Resilience (Australian Government, 2011), EMA advocates the *Prevent, Prepare, Respond, Recover* (PPRR) model first developed in the US and used in Australia since 1989 (Cronstedt, 2002, EMA, 2004). This model has been subject to significant criticism,[18] particularly in light of its apparent shortcomings when managing a number of recent climate disasters (Heazle et al., 2013; Cronstedt, 2002). Nonetheless, the disaster risk and emergency management provisions in Queensland have been commended by an independent Commission of Inquiry in the aftermath of the Brisbane floods of 2010/2011 (QFCI, 2012, p. 30), and across Australia more broadly by the US Federal Emergency Management Agency (FEMA), particularly in comparison to similar provisions in other parts of the world (Peters and McEntire, 2014). Indeed, FEMA note that Queensland in particular 'is one of the most prepared states in Australia' (Peters and McEntire, 2014: p. 22).

In Queensland, the Disaster Risk Management Act (2003) is the principal legislative mechanism driving climate adaptation policy and practice. The Act mandates the formation of disaster management groups at state, district and local level and the preparation of disaster risk management plans and guidelines (Bajracharya, 2011). Queensland's disaster risk management provisions are highly decentralised, taking a bottom–up approach but with strategic oversight at state government level (Peters and McEntire, 2014). Local governments are responsible for identifying and evaluating hazards and risks in collaboration with the emergency services, voluntary interest groups and community organisations. This is achieved through local disaster management groups and their derivation of local disaster management plans to enhance communities' preparedness for, and response to, extreme events and thus, their climate resilience.

There are, of course, similarly structured emergency management arrangements in the UK (e.g., through the Civil Contingencies Act (2004)) (UK Government, 2013). However, the UK's lack of experience with, or exposure to, frequent and contrasting extreme events must be considered in relation to managing climate change. The efficacy of Queensland's disaster risk management provisions for enhancing climate change resilience can, perhaps, be best summarised by the adage 'exposure breeds resilience'; on which basis, I argue, this region holds a considerable advantage:

> Australia has come a long way in how it responds to unexpected events, whether they be natural or man–made. Today, the emergency management sector is like a well–oiled machine.
>
> (Jones, 2007; cited in Peters and McEntire, 2014: p. 29)

Importantly, the policy and planning issues being addressed under the headings of emergency and disaster risk management, and urban or land–use planning in Queensland are many of the same issues that are specifically branded as Climate Change Adaptation in other countries, and in particular in the UK. Such examples include the provision of flood risk management assets (such as levees and barriers) and preparation and adaptation guidelines for local government planning (see for example, EA, 2010a,b). What remains in question, therefore, is the extent to which the climate adaptation actions implemented under Queensland's alternative policy brandings are less effective due to a recurring failure to consider the outputs of scientific projections of future climate change.

Although the provision of public assets and infrastructure, the development of planning guidelines and the management of agriculture may be less effective over the long term by failing to consider the potential impacts from climate change at the current time, policymaking efforts toward enhancing community preparedness, response and adaptive capacity, that are a necessary priority of Queensland's climate adaptation efforts in the face of persistent and unpredictable extreme events, are not disadvantaged in the same way. These provisions, I argue, are equally important components in the development of effective climate change adaptation. If, as I argue in Chapter 8, climate resilience can be best achieved by balancing both the *flexibility* and *stability* of communities and government, then the development of communities' social capital or adaptive capacity – associated with effectively preparing for, managing and responding to climate variability and both chronic and acute climate extremes, is as valuable a form of climate change adaptation as 'predict-then-act' climate risk management methods often associated with infrastructure provision and planning (Heazle et al., 2013). On this basis, it seems fair to conclude that Queensland's efforts to adapt to existing climate extremes has ensured (somewhat unintentionally) significant progress toward climate *change* adaptation which, over the medium term at least, is comparable to adaptation efforts in the UK. Indeed, in terms of the development of local government and community awareness and resilience to extreme events and the management of emergency services, Queensland appears to be considerably more advanced than the UK.

The UK

The UK is generally seen as a world leader in the field of climate change adaptation policy (Keskitalo, 2010) a status that appears to be largely based upon their enthusiasm for the development of useful climate change science and its use for impact and risk assessments. Efforts in this regard began in the 1990s when the government first commissioned climate change scenarios to inform policymaking. Two iterations of scenarios produced by a government-sponsored Climate Change Impacts Review Group (CCIRG) were followed by the establishment in 1997 of the UK Climate Impacts Program (UKCIP), a boundary organisation (Guston, 2001) assigned the task of making climate change science outputs relevant for government, business and public users (Hulme and Dessai, 2008; McKenzie Hedger et al., 2000).

UKCIP has been at the forefront of climate change adaptation policymaking since its inception, reflecting both the legitimacy given to climate science by successive UK governments and prevailing ideals concerning its use to inform decision-making in a linear and instrumental way. One of the principal advisees of UKCIP, in theory at least, has been local government, even if, in recent years they have had neither the resources nor the policy framework to utilise this evidence in practice (see Chapter 6). Between 1997 and 2010, under the Blair and Brown governments, climate change was a significant component of the governance provisions of local authorities (DEFRA, 2010). Nonetheless, the coalition government under David Cameron subsequently removed these provisions, reflecting its political priority to reduce burdens on the state and to increase the independence of local government (Taylor-Gooby, 2012). At the same time, Cameron established a national framework on community resilience for local governments, thereby fostering neo-liberal governance ideals that, in tandem with the Civil Contingencies Act (2004) governing emergency management provisions, places principal adaptation policy responsibility onto local communities (Joseph, 2013).

At national level, a UK Climate Change Programme was first put in place by the Blair government in 2000, and updated in 2006, with provisions to address both mitigation and adaptation (DETR, 2000; DEFRA, 2006). Much of the content of this policy framework was subsequently channelled into the provisions of the Climate Change Act (2008) (CCA) (Keskitalo, 2010). Under the CCA, government is obliged to assess the risks from climate change and to develop and implement a national climate change adaptation plan, to be revised at five-yearly intervals. It provides governments with the power to oblige key public service and infrastructure providers to undertake a similar assessment of risks and to develop adaptation plans every five years. In addition, the Committee on Climate Change – the organisation responsible for monitoring governments' progress on meeting its commitments – has established a subcommittee to assess government's progress on adaptation. The Environment Agency,[19] similarly, began considering climate change adaptation in their responsibilities to government in 2005, and since 2008 the Agency has had a climate change strategy as well as a series of adaptation plans covering the major themes of its responsibilities to government (EA, 2008) (see Chapter 8), now incorporated into their reporting requirements under the CCA (EA 2010a).[20]

Under the aforementioned obligations of the CCA, government commissioned its first UK Climate Change Risk Assessment (UKCCRA) in 2012 (see Chapter 7). The assessment involved a number of stages of analysis, beginning with an initial assessment of 700 risks, with more in-depth analysis of a short-list of 100 risks (DEFRA, 2012c). Further to this assessment, the government subsequently published a National Adaptation Programme (NAP) that was initially intended to address the risks outlined in the UKCCRA (2012),[21] alongside separate programmes relating to matters devolved to Scotland, Wales and Northern Ireland (DEFRA, 2013a). A second UKCCRA (2017) has recently been published (ASC, 2016); this time a considerably more streamlined assessment of 60 risks.

The shorter format of the UKCCRA (2017), and its focus on those risks considered to be urgent, may reflect Tory governments' considerably diminished interest in adaptation activities relative to their Labour predecessors, as they have been clear that it will only address adaptation where there is a clear economic signal demonstrating a need for action (UKCIP, pers comm, 2013; Climate UK, pers comm, 2013). As significantly perhaps, has been the dismantling of adaptation reporting provisions for local government and a downgrading of previous obligations for key service and infrastructure providers to report their climate risks, now to be completed on a voluntary basis only (DEFRA, 2014). As such, the legal obligations on sub-national government or industry to explicitly address climate change adaptation have been considerably diminished, while a renewed emphasis has been placed on the need for a cost-benefit case for adaptation investment.

There are, however, a variety of other significant policy levers and drivers of adaptation at national, regional and local levels. These primarily originate from the Environment Agency and the Water Services Regulation Authority (OFWAT) that continue to incorporate climate change adaptation into their regulatory and policy implementation activities (OFWAT, 2010; EA 2013) as well as in the provision of water resource and flood risk management infrastructure (EA, 2010a). Indeed, these latter provisions appear to be some of the strongest attributes of the UK's adaptation to future climate change (see Box. 4.3, Chapter 4). The Cameron government's revised planning policy regime retained a requirement for local government to address climate change adaptation even though there are no longer any reporting requirements to demonstrate such policy action (DCLG, 2012: p. 23). Furthermore, there is a range of other government-funded organisations tasked with the consideration of climate change adaptation. The first such organisation was UKCIP, as discussed above, which has now been replaced in its public advisory role by the Environment Agency's Climate Ready support service. Alongside this service, Climate UK – a public-private voluntary partnership – was established to maintain a network of regional stakeholder engagement forums to consider climate change at local government level. Unlike the comparative case of Queensland, however, the priority focus of most adaptation initiatives in the UK to date has been upon understanding future climate change, with very limited experience of managing or coordinating responses to extreme climate conditions or events (see for instance, DEFRA, 2013a; EA, 2010a).

Adaptation plans developed to date by national government, the Environment Agency and other government institutions have focused on the reduction of uncertainty relating to potential climate change impacts, while most adaptation-related provisions have actually occurred under the guise of concurrent policy priorities such as flood risk, water resources management and urban planning (ASC, 2016; Tompkins et al., 2010). These policy provisions have often been relabelled as climate change adaptation actions under the CCA, at least when it has been convenient to do so. Through the establishment of organisations such as the Climate Ready support service, its predecessor UKCIP, and voluntary initiatives such as Climate UK, as well as the reporting and evidence development provisions of the CCA, the UK's tangible efforts toward climate *change* adaptation have principally been focused on ensuring sufficient capacity is built into existing infrastructure provisions and developing a policy framework between levels of government to understand and integrate the threats associated with potential future climate change (Tompkins et al., 2010). Outside of the remit of the CCA, however, and particularly at local government level, adaptation appears to have become an increasingly lower priority. By contrast, while the Australian case has paid scant attention to climate change, the extant hazards faced on a year-to-year basis have ensured the maintenance of tried and tested policies for climate adaptation including a world-class community and local government adaptive capacity framework to manage extreme climate conditions (Peters and McEntire, 2014; Dedekorkut-Howes et al., 2010).

Ultimately, I argue, questions remain about how effective the UK's adaptation provisions will actually be in practice, considering the country's lack of practical experience in dealing with climate extremes and their relative dearth of local government provisions and experience. Moreover, and as described in detail in Chapter 6, the UK approach to adaptation has placed considerably greater emphasis on the use and usability of climate change science for informing policy in a linear way than the comparable case of Queensland, with only limited demonstration that this enhanced understanding holds tangible benefits. The UK's emphasis on more and better science reflects the greater legitimacy granted to climate change as a policy issue in the UK, however it does not necessarily translate into more evidence-based decision-making. Nor does it necessarily translate into enhanced climate resilience.

I argue that the legitimacy of climate science in the UK has, to date, been of limited use when pre-emptive adaptation investments must be justified on the basis of a 'business case'. This is because adaptation investments that could be justified on the basis of climate change projections, can in practice only be made where such initiatives pass the cost-benefit test; a mechanism which heavily favours present concerns over longer term risks due to the discount rates used, and the lack of certainty associated with the science of, and therefore the long term costs from, future climate change.

As I describe in later chapters, climate scientists cannot predict the future and the inappropriate use of climate science may even result in maladaptation through strongly path-dependent policy initiatives. The UK approach, therefore, sits in stark contrast to that of Queensland; the latter progressing tangible adaptation policy without significant recourse to climate science, by focusing upon the present resilience of communities to extreme events.

Conclusion

The analysis provided in this chapter suggests that the legitimacy of different types of evidence for climate adaptation policy is related to a broader legitimacy for the task of managing climate change. Where climate change is considered a worthy political priority, climate science is granted legitimacy and by association the task of climate change adaptation is embraced. Where climate change is not legitimate, climate science is abandoned, and so too, the task of climate *change* adaptation. As discussed further in later chapters, however, in both Queensland and the UK, science is still superseded by evidence relating to the short-term costs and benefits of investment decisions. Both case-study regions have a history of legitimising and using evidence for environmental policymaking only to the extent that evidence is congruent with the socio-economic and cultural norms and priorities of both the public and its political executive.

In Queensland, these norms and priorities have ensured that climate change science has remained a contested issue. Although anthropogenic climate change may be accepted as scientifically credible,[22] the use of climate science for policymaking conflicts with foundational socio-economic assumptions in the state in relation to the role of government and the exploitation of natural resources. Conversely, a policy focus on climate exposure and resilience not only reflects the nature of extant climate extremes, it also plays to popular expectations that society must battle with the natural environment and eliminate environmental hazards. In the UK too, environmental policy has a history of only relying on evidence where it is politically expedient to do so. Therefore, just as the issue of acid rain was a victim of this selective recourse to the facts, the issue of climate change and adaptation has been a beneficiary. Climate change and the use of climate science align with fundamental historical and socio-economic values and priorities relating to energy independence, a love of the environment and conservation amongst a large proportion of the population, and a desire for continuing leadership amongst the international community, despite its contemporary diminutive stature as a sovereign power.

In terms of tangible policy outcomes, however, I conclude that the cases of Queensland and the UK are largely equivalent. The UK can certainly be considered more progressive in its evidence-based values, as it focuses its attention on the development of usable climate science and institutional structures that can account for the future risks from climate change. Yet, these aspirations are nonetheless stymied by priorities for balancing short-term costs and benefits when considering infrastructure investment, and can result in policy paralysis until the uncertainties about future climate change have been appropriately reduced (assuming that they can, in fact, be reduced). By contrast, Queensland has the benefit of a tried and tested policy framework that continues to maintain the region's resilience to extreme weather and climate events. In both cases climate *change* adaptation lacks legitimacy at local government level, and in both cases there is a dearth of tangible climate change-specific infrastructure or service provision. This comparability of policy outcomes between cases reveals an interesting point about the use of evidence for adaptation policy. While political acceptability for climate change is a necessary pre-determinant of the legitimacy of climate science for policymaking, the existence (or otherwise) of legitimacy for climate science may have less impact upon the outcomes of adaptation policymaking than is often assumed, in as far as it enhances society's climate resilience or reduces its vulnerability, at least over the medium term.

As I demonstrate in Chapters 6 and 7, however, what legitimacy for climate science can do is influence perceptions about the credibility and salience of this evidence. Moreover, when this legitimacy is framed in linear-technocratic terms it can facilitate evidence politicisation, and thus foment the scientisation of adaptation policymaking. This conclusion about climate science suggests that such complex, uncertain and contentious evidence has limited utility for policymaking in practice, at least in the ways expected under the linear-technocratic model.

Notes

1   Unlike under the evidence-based provisions of the Climate Change Act, adaptation policy at local government level in the UK is considerably less well developed and in recent years, has increasingly been subsumed under policy frameworks concerned with climate resilience (see for example, UK Government, 2014).
2   Parts of this comparative analysis were first published in abridged form in Tangney and Howes (2016).
3   At the time of completing the final draft of this book, the UK has decided to leave the European Union, as allowed under the terms of the Lisbon Treaty, which was signed

by EU member states in 2007. This unexpected shift in European political order provides a fascinating twist to the international politics of climate change and will undoubtedly influence the UK's pursuit of climate change policy. It does not, however, invalidate the conclusions of the comparative analysis presented here. The implications of 'Brexit' are uncertain at this time and as described here, many of the substantive policy drivers for climate change adaptation come from the UK's own Climate Change Act (2008) rather than directly from Europe.

4   A deliberate misnomer by Abbott; the Clean Energy Act (2011) incorporated an ETS, not a tax, with a fixed price for the first three years.

5   Notably, only five MPs voted against it.

6   Mirroring the challenges faced by Malcolm Turnbull as leader of the LNP coalition in Australia.

7   A declaration, signed by the three main political party leaders in 2015, stated: 'Climate change is one of the most serious threats facing the world today. It is not just a threat to the environment, but also to our national and global security, to poverty eradication and economic prosperity … Acting on climate change is also an opportunity for the UK to grow a stronger economy, which is more efficient and more resilient to the risks ahead' (Carrington, 2015).

8   Despite considerable efforts at better integration. For example, moves towards cooperation by the Council of Australian Governments on climate change adaptation (COAG, 2007) and DRM (COAG, 2011).

9   The practice of shifting policy responsibility to lower levels of governance without the corresponding devolvement of funds necessary for effective policy response.

10   '[A] theory of political economic practices that proposes that human well-being can best be advanced by liberating individual entrepreneurial freedoms and skills within an institutional framework characterized by strong private property rights, free markets, and free trade' (Harvey, 2007, p. 2). The interventionist governance paradigm was prevalent for all areas of environmental management across Australia until the 1980s, at which point 'neo-liberal' policies caused a shift away from previous priorities of state protectionism, with a corresponding increase in reliance on the distributive powers of the market (Pusey, 1991). This trend was prompted by prevailing forces of globalisation worldwide (Harvey, 2007) and, to some extent, the pressure on Australian government for economic integration with its international trading partners (Pusey, 1991). While market-based strategies for addressing climate change are most likely here to stay, and were explicitly advocated by Professor Ross Garnaut in his high-profile Climate Change Review to federal government in 2008, the review also recognised the limitations of these approaches and the need for government intervention for climate adaptation (Curran, 2011; Garnaut, 2008: p. xxxii, xi).

11   A term used in economics to describe a situation in which inflation and unemployment rates remain high, while the rate of economic growth slows down.

12   As demonstrated, for example, in 2010 by the Cameron government's repeal of the statutory reporting requirement on state infrastructure and public service providers under the Climate Change Act (2008), in favour of a voluntary reporting scheme.

13   Organisation of the Petroleum Exporting Countries (OPEC): a consortium of 12 oil producing countries from the Middle East, Africa and South America that coordinates the pricing and production policies of its members' oil supplies.

14   I was one of the authors of the UK government's final regulatory impact assessment for the Water Framework Directive (DEFRA, 2007).

15   This interesting tactic of invoking the need for more and better information also mirrors similar commitments in the US, used as a means to avoid more substantive policy measures that would be politically divisive (Grundmann, 2006).

16   The Council of Australian Governments (COAG) is the principal intergovernmental forum of Australia. The council comprises the Prime Minister, State and Territory Premiers and Chief Ministers as well as the President of the Australian Local

Government Association: 'The role of COAG is to promote policy reforms that are of national significance, or which need co-ordinated action by all Australian governments' (COAG, 2015).

17 See Chapter 8 for a discussion of the varying concepts of climate resilience.

18 This criticism should be taken in the context of understanding the preoccupation of Australia's governments and research institutions with managing climate extremes, relative to other countries. As a very loose indication, a cursory Google Scholar search at the time of writing, using the search terms (Australia OR Queensland; 'Disaster Risk' OR 'Emergency Management') yields 16,800 hits. An equivalent UK search (UK OR England OR Britain; 'Disaster Risk' OR 'Emergency Management') yields 17,200 hits. Given that both the population and research community of Australia is roughly one-third of the size of the UK, this contrast appears to point to Australia's historical preoccupation with climate risk management.

19 The Environment Agency of England & Wales is the principal institution responsible for the regulation and implementation of environmental policy in the UK, alongside the Scottish Environmental Protection Agency and the Northern Ireland Environment Agency.

20 I was the lead author of the Environment Agency's Marine Climate Change Adaptation Plan (2009), as well as principal author of the Environment Agency's report to government required under the terms of the Climate Change Act (2008) (EA, 2010a).

21 As discussed in Chapter 7, in practice the links between the UKCCRA 2012 and the National Adaptation Programme 2013 appear to have been influenced by political machinations.

22 Formally at least, though not necessarily in open political or parliamentary debate.

Bibliography

Abbott, T 2013, '1st Leaders Debate for the 2013 Federal Election', opening speech broadcast on 11 August, viewed 13 August 2013, www.abc.net.au/iview/#/series/12771/.

ABC 2014a, 'Direct action: Government releases policy white paper on climate change plan', Australian Broadcasting Corporation, 25 April, viewed 17 June 2014, www.abc.net.au/news/2014-04-24/government-releases-climate-change-policy-white-paper/5409262/.

ABC 2014b, 'As it happened: Senate passes legislation to repeal carbon tax', Australian Broadcasting Corporation, 17 July, viewed 17 June 2014, www.abc.net.au/news/2014-07-17/live-blog-coalition-in-bid-to-push-through-carbon-tax-repeal/5603830/.

ABC 2014c, 'Jeff Seeney orders Moreton Bay Regional Council to remove references to climate change-derived sea level rises from regional plan', Australian Broadcasting Corporation, 9 December, viewed 4 February 2015, www.abc.net.au/news/2014-12-09/seeney-removes-climate-change-references-from-council-plan/5954914/.

ABC 2015, 'Jeff Seeney said climate change 'semi-religious belief': Queensland mayor signs statutory declaration stating Deputy Premier made comment', Australian Broadcasting Corporation, 23 January, viewed 4 February 2015, www.abc.net.au/news/2015-01-23/jeff-seeney-denies-he-said-climate-change-was-a-semi-religious-/6041710/.

Adaptation Sub-Committee [ASC] 2016, 'UK Climate Change Risk Assessment 2017 Synthesis Report: Priorities for the next five years', Adaptation Sub-Committee of the Committee on Climate Change, London.

Anderson, K, Bows, A and Mander, S 2008, 'From long-term targets to cumulative emissions pathways: Reframing UK climate policy', *Energy Policy*, vol. 36, no. 10, pp. 3714–3722.

Assinder, N 2000, 'Blair seeks to woo green vote', BBC News Online, 24 Oct, viewed 15 December 2014, http://news.bbc.co.uk/2/hi/uk_news/politics/988323.stm/.

Australian Bureau of Statistics 2012, 'Population estimates by local government area, 2001–2011', viewed 13 August 2013, www.abs.gov.au/AUSSTATS/abs@.nsf/DetailsPage/3218.02011/.

Australian Government 2011, *National Strategy for Disaster Resilience*, Commonwealth of Australia, viewed 2 January 2017, www.ag.gov.au/EmergencyManagement/Emergency-Management-Australia/Pages/default.aspx/.

Australian Government 2014a, *Climate Change Adaptation Program: Department of Environment*, Commonwealth of Australia, viewed 19 July 2014, http://www.environment.gov.au/climate-change/adaptation/climate-change-adaptation-program./

Australian Government 2016, *Emergency Management Australia*, Australian Government website, viewed 2 January 2017, www.ag.gov.au/EmergencyManagement/Emergency-Management-Australia/Pages/default.aspx/.

Australian Institute for Disaster Resilience [AIDR] 2016, website viewed 2 January 2017, www.aidr.org.au/.

Bajracharya, B, Childs, I and Hastings, P 2011, 'Climate change adaptation through land use planning and disaster management: Local government perspectives from Queensland', Paper presented to the 17th Pacific Rim Real Estate Society Conference, 16–19 January 2011, viewed 31 March 2017, *www*.prres.net/papers/Bajracharya_Childs_Hastings_Climate_change_disaster_management_and_land_use_planning.pdf.

Bayley, CA 2004, *The Birth of the Modern World, 1740–1914: Global Connections and Comparisons*, Blackwell, London.

BBC 2004, 'The Coal Strike 1984–1985', British Broadcasting Corporation, March 2004, viewed 17 September 2013, www.bbc.co.uk/bradford/sense_of_place/miners/miners_strike.shtml/.

Beeson, M and McDonald, M 2013, 'The politics of climate change in Australia', *Australian Journal of Politics and History*, vol. 79, no. 3, pp. 331–348.

Bell, S 2006, 'Concerned scientists, pragmatic politics and Australia's green drought', *Science and Public Policy*, vol. 33, no. 8, pp. 561–570.

Benedick, RE 2001, 'Striking a new deal on climate change', *Issues in Science and Technology*, vol. 18, no. 1, pp. 71–77.

Brisbane City Council [BCC] 2007, *Brisbane's Plan for Action on Climate Change and Energy 2007*, Brisbane City Council, viewed 22 July 2014, http://www.brisbane.qld.gov.au/sites/default/files/20140414%20-%20Brisbanes%20Plan%20for%20Action%20on%20Climate%20Change.pdf.

Bulkeley, H 2001, 'No regrets? Economy and environment in Australia's domestic climate change policy process', *Global Environmental Change*, vol. 11, no. 2, pp. 155–169.

Bureau of Meteorology [BoM] 2016, 'Tropical Cyclones in Queensland', website viewed 16 December 2016, www.bom.gov.au/cyclone/about/eastern.shtml#history/.

Carrington, D 2015, 'Cameron, Clegg and Miliband sign joint climate change pledge', *The Guardian*, 14 February, viewed 19 March 2015, www.theguardian.com/environment/2015/feb/14/cameron-clegg-and-miliband-sign-joint-climate-pledge/.

Carter, N 2009, 'Vote blue, go green? Cameron's Conservatives and the environment', *The Political Quarterly*, vol. 80, no. 2, pp. 233–242.

Christoff, P 1998, 'From global citizen to renegade state: Australia at Kyoto', *Arena*, vol. 10, pp. 113–127.

Christoff, P 2005, 'We're all greenies now? Recent patterns in environment policy in Australia and New Zealand', *Around the Globe*, vol. 2, no. 2, pp. 6–9.

Christoff, P 2010, 'Touching the void: The Garnaut Review in the chasm between science, economics and politics', *Global Environmental Change*, vol. 20, pp. 214–217.

Christoff, P 2008, 'Aiming high: On Australia's emissions reductions targets', *UNSW Law Journal*, vol. 31, no. 3, pp. 861–879.

Climate Institute, The 2015, *Climate of the Nation 2015: Australian Attitudes on Climate Change*, The Climate Institute website, viewed 6 January 2016, www.climateinstitute. org.au/verve/_resources/Climate_of_the_Nation_web_final.pdf.

Council of Australian Governments [COAG] 2007, *National Climate Change Adaptation Framework, Council of Australian Governments*, Commonwealth of Australia 2010, viewed 09 September 2012: www.climatechange.gov.au/government/initiatives/national-climate-change-adaptation-framework.aspx/.

Council of Australian Governments [COAG] 2011, *National Strategy for Disaster Resilience: Building Our Nation's Resilience to Disasters*, NEMC Working Group tasked by the COAG, viewed 16 August 2013: www.coag.gov.au/.

Cronstedt, M 2002, 'Prevention, preparedness, response, recovery: AN outdated concept?', *Australian Journal of Emergency Management*, vol. 17, no. 7, pp. 10–13.

Crowley, K 2007, 'Is Australia faking it? The Kyoto Protocol and the greenhouse policy challenge', *Global Environmental Politics*, vol. 7, no. 4, pp. 118–139.

Crowley, K and Walker, KJ (eds) 2012, *Environmental Policy Failure: The Australian Story*, Tilde University Press, Victoria, Australia.

Curran, G 2009, 'Ecological modernisation and climate change in Australia', *Environmental Politics* vol. 18, no. 2, pp. 201–217.

Curran, G 2011, 'Modernising climate policy in Australia: Climate narratives and the undoing of a prime minister', *Environment and Planning C: Government and Policy*, vol. 29, no. 6, pp. 1004–1017.

Damro, C and Mendes, PL 2005, 'Emissions Trading at Kyoto: From EU Resistance to Union Innovation', in Jordan, A (ed.), *Environmental Policy in the European Union: Actors, Institutions and Processes*, 2nd edn, Earthscan, London.

Dedekorkut-Howes, A and Howes, M 2014, 'Climate Adaptation Policy and Planning in South East Queensland', in Burton, P (ed.), *Responding to Climate Change: Lessons from a Hotspot*, CSIRO Publishing, Melbourne.

Dedekorkut, A, Mustelin, J, Howes, M and Byrne, J 2010, 'Tempering growth: Planning for the challenges of climate change and growth management in SEQ', *Australian Planner*, vol. 47, no. 3, pp. 203–215.

Department of Climate Change [DCC], 2010, *Adapting to Climate Change in Australia: An Australian Government Position Paper*, DCC, Canberra, viewed 31 March 2017, http://coastaladaptationresources.org/PDF-files/1236-gov-adapt-climate-change-position-paper.pdf.

Department of Communities and Local Government [DCLG] 2012, *National Planning Policy Framework: Department of Communities and Local Government*, March 2010, viewed 30 October 2013, https://www.gov.uk/government/uploads/system/uploads/attachment_data/file/6077/2116950.pdf.

Department of Environment, Food & Rural Affairs [DEFRA] 2007, *Overall Impact Assessment of the Water Framework Directive (2000/60/EC)*, viewed 1 February 2015, http://archive.defra.gov.uk/environment/quality/water/wfd/documents/RIA-river-basin-v2.pdf.

Department of Environment, Food & Rural Affairs [DEFRA] 2010, *National Indicator 188: Adapting to Climate Change*, 6 April 2010, *viewed* 28 October 2013, http://archive.defra.gov.uk/corporate/about/with/localgov/indicators/ni188.htm./

Department of Environment, Food & Rural Affairs [DEFRA] 2012c, *Climate Change Risk Assessment Methodology Report,* July 2012, *viewed* 31 March 2017, http://randd.defra.gov.uk/Default.aspx?Module=More&Location=None&ProjectID=15747./

Department of Environment, Food & Rural Affairs [DEFRA] 2013a, *The National Adaptation Programme: Making the Country Resilient to Changing Climate,* HMSO, London, *viewed* 31 March 2017, https://www.gov.uk/government/publications/adapting-to-climate-change-national-adaptation-programme./

Department of Environment, Food & Rural Affairs [DEFRA] 2014, *Adapting to Climate Change,* HM Government 2014, *viewed* 21 July 2014, https://www.gov.uk/government/policies/adapting-to-climate-change./

Department of Environment, Transport and the Regions [DETR] 2000, *Climate Change: The UK Programme,* HMSO, Norwich, UK, *viewed* 31 March 2017, www.cne-siar.gov.uk/emergencyplanning/documents/Climate%20Change%20-%20UK%20Programme.pdf.

Devine M., 2002. 'Drought relief sceptics must be hung out to dry'. *The Sun-Herald,* 13 October 2002, viewed 2 August 2013, www.smh.com.au/articles/2002/10/12/1034222635512.html./

Dovers, S (ed.) 2000, *Environmental History and Policy: Still Settling Australia.* Oxford University Press, Melbourne, Victoria, Australia.

Dunn, B 2014, 'Making sense of austerity: The rationality in an irrational system', *The Economic and Labour Relations Review,* vol. 25, no. 3, pp. 417–434.

Emergency Management Australia [EMA] 2004, *Emergency Management in Australia: Concepts and Principles,* Attorney General's Department, Commonwealth of Australia, Canberra, viewed online 17 March 2015, www.em.gov.au/Documents/Manual01-EmergencyManagementinAustralia-ConceptsandPrinciples.pdf .

Environment Agency of England & Wales [EA] 2008, *Climate Change Adaptation Strategy 2008–11,* EA, Bristol, UK, viewed 30 October 2013, www.environment-agency.gov.uk/static/documents/Leisure/adaptation_strategy_2083410.pdf.

Environment Agency of England & Wales [EA] 2010a, *Managing the Environment in a Changing Climate,* EA, Bristol, UK, viewed 17 March 2015, www.environment-agency.gov.uk/research/library/publications/130528.aspx/.

Environment Agency of England & Wales [EA] 2010b, *State of the Environment–South East England,* EA, Bristol, UK, viewed 20 August 2013, www.environment-agency.gov.uk/research/library/publications/34103.aspx/.

Evans, D 2003, *A History of Nature Conservation in Britain,* 2nd edn, Routledge, London.

Evans, R 2007, *A History of Queensland,* Cambridge University Press, Cambridge.

Fitzgerald, R 1984, *A History of Queensland from 1915 to the 1980s,* University of Queensland Press, St Lucia, Queensland, Australia.

Flannery, T 1994, *The Future Eaters: An Ecological History of the Australasian Lands and People,* New Holland Publishers, Chatswood, NSW.

Garnaut, R 2008, *Garnaut Climate Change Review: Final Report,* Cambridge University Press, Cambridge.

Garner, R 2000, *Environmental Politics: Britain, Europe and the Global Environment,* 2nd edn, Macmillan Press, London.

Gold Coast City Council [GCCC] 2009, *Climate Change Strategy 2009 – 2014,* GCCC, viewed 22 January 2014, www.goldcoast.qld.gov.au/documents/bf/climate_strategy.pdf.

Gray, TS 1997, 'Politics and the Environment in the UK and Beyond', in Redclift, M and Woodgate, G (eds), *The International Handbook of Environmental Sociology*, Edward Elgar Publishing, Cheltenham, UK.

Grice, A 2009, 'Cameron hit by Tory backlash on environment', *The Independent*, 2 December, viewed 24 October 2013, www.independent.co.uk/news/uk/politics/ cameron-hit-by-tory-backlash-on-environment-1832208.html/.

Griffiths, A, Haigh, N and Rassias, J 2007, 'A framework for understanding institutional governance systems and climate change: The case of Australia', *European Management Journal*, vol. 25, no. 6, pp. 415–427.

Grubb, M 2002, 'Britannia waives the rules: The UK, the EU and climate change', *New Economy*, vol. 9, no. 3, pp. 139–142.

Grundmann, R 2006, 'Ozone and climate: Scientific consensus and leadership', *Science, Technology & Human Values*, vol. 31, no. 1, pp. 73–101.

Guston, DH 2001, 'Boundary organizations in environmental policy and science: An introduction', *Science Technology, & Human Values,* vol. 26, no. 4, pp. 299–408.

Harling, P 2001, *The Modern British State: An Historical Introduction*, Polity Press, Cambridge.

Hajer, M 1995, *The Politics of Environmental Discourse: Ecological Modernisation and the Policy Process*, Oxford University Press, Oxford.

Harvey, D 2007, *A Brief History of Neoliberalism*, Oxford University Press, Oxford.

Head, B 2008b, 'Three lenses of evidence-based policy', *Australian Journal of Public Administration*, vol. 67, no. 1, pp. 1 – 11.

Head, L, Adams, M, McGregor, HV and Toole, S 2014, 'Climate change and Australia', *WIRES Climate Change*, vol. 5, no. 2, pp. 175–197.

Heazle, M, Tangney, P, Burton, P, Howes, M, Grant-Smith, D, Reis, K, Bosomworth, K 2013, 'Mainstreaming climate change adaptation: An incremental approach to disaster risk management in Australia', *Environmental Science & Policy*, vol. 33, pp. 162–170.

Hennessy, K, Fitzharris, B, Bates, BC, Harvey, N, Howden, SM, Hughes, L, Salinger, J and Warrick, R 2007, 'Australia and New Zealand', in Parry, ML, Canziani, OF, Palutikof, JP, van der Linden, PJ and Hanson, CE (eds), *Climate Change 2007: Impacts, Adaptation and Vulnerability. Contribution of Working Group II to the Fourth Assessment Report of the Intergovernmental Panel on Climate Change*, Cambridge University Press, Cambridge.

Herbohn, K, Dargusch, P and Herbohn, J 2012, 'Climate change policy in Australia: Organisational Responses and Influences', *Australian Accounting Review*, vol. 22, no. 2, pp. 208–222.

Howes, M 2008, 'Rethinking governance: Lessons in collaboration from environmental policy', Australasian Political Studies Association Annual Conference 2008, 6–8 July, viewed 13 August 2013, www98.griffith.edu.au/dspace/handle/10072/22581/.

Howes, M and Dedekorkut-Howes, A 2010, 'From white shoes to waders: Climate change adaptation and government on the Gold Coast', Australian Political Science Association Conference 2010, viewed 5 August 2013, http://apsa2010.com.au/full-papers/pdf/ APSA2010_0056.pdf.

Howes, M, Grant-Smith, D, Bosomworth, K, Reis, K, Tangney, P, Heazle, M, McEvoy, D and Burton, P 2012, *The Challenge of Integrating Climate Change Adaptation and Disaster Risk Management: Lessons from Bushfire and Flood Inquiries in an Australian Context*, Urban Research Program, Issues Paper 17, Griffith University, Brisbane, Australia.

Howes, M, Grant-Smith, D, Reis, K, Bosomworth, K, Tangney, P, Heazle, M, McEvoy, D and Burton, P 2013, *Rethinking Disaster Risk Management and Climate Change Adaptation*, National Climate Change Adaptation Research Facility, Gold Coast, pp. 63, viewed

5 August 2013, http://apo.org.au/sites/default/files/docs/Howes-2013-rethinking-disaster-risk-mngt-WEB.pdf.

Howes, M, Tangney, P, Reis, K, Grant-Smith, D, Heazle, M, Bosomworth, K and Burton, P 2015, 'Towards networked governance: Improving interagency communication and collaboration for disaster risk management and climate change adaptation in Australia', *Journal of Environmental Planning and Management*, vol. 58, no. 5, pp. 757–776.

Hudson, M 2016, 'The sound of silence: Why has the environment vanished from election politics?' *The Conversation*, 23 June, viewed 28 June 2016, https://theconversation.com/the-sound-of-silence-why-has-the-environment-vanished-from-election-politics-59658.

Hulme, M and Dessai, S 2008, 'Negotiating future climates for public policy: A critical assessment of the development of climate scenarios for the UK', *Environmental Science & Policy*, vol. 11, pp. 54–70.

Johnston, WR 1988, *A Documentary History of Queensland*, University of Queensland Press, St Lucia, Queensland.

Jones, R 2007, 'A Nation Comes Together: A Few Pivotal Events Changed the Way Australia Responds to its Crises', in Keeney, J (ed.), *In Case of Emergency: How Australia Deals with Disasters and the People who Confront the Unexpected*. Design Master Press, NSW, Australia.

Jordan, A 2001, 'National environmental ministries: managers or ciphers of European Union environmental policy?', *Public Administration*, vol. 79, no. 3, pp. 643–664.

Jordan, A 2002a, 'European Union Environmental Policy and Britain', in Jordan, A (ed.), *The Europeanization of British Environmental Policy: A Departmental Perspective*, Palgrave Macmillan, Basingstoke.

Jordan, A 2002b, 'Efficient Hardware and Light Green Software: Environmental Policy Integration in the UK', in Lenschow, A (ed.), *Environmental Policy Integration: Greening Sectoral Policies in Europe*, Earthscan, London.

Jordan, A (ed) 2005, *Environmental Policy in the European Union: Actors, Institutions and Processes*, 2nd edn, Earthscan, London.

Joseph, J 2013, 'Resilience as embedded neoliberalism: A governmentality approach', *Resilience*, vol. 1, no. 1, pp. 38–52.

Keskitalo, E and Carina, H 2010, 'Climate Change Adaptation in the United Kingdom: England and Southeast England', in Brannlund, L (ed.), 2010, *Developing Adaptation Policy and Practice in Europe: Multi-Level Governance of Climate Change*, Dordrecht, Heidelberg, London and New York, Springer Science and Business Media.

Lane, JE 2000, *New Public Management*, Routledge, London, UK.

Lean, G 1998, 'How Britain came clean', *The Independent*, 5 July 1998, London.

Lowe, P and Ward, S (eds) 1998, *British Environmental Policy and Europe: Politics and Policy in Transition*, Routledge, London.

McCormick, J 2002, 'Environmental Policy in Britain', in Desai, U (ed.), 2002, *Environmental Politics and Policy in Industrialized Countries*, MIT Press, Cambridge, MA.

McDonald, J, Baum, S, Crick, F, Czarnecki, J, Field, G, Low Choy, D, Mustelin, J, Sano, M and Serrao-Neumann, S 2010, 'Climate change adaptation in South East Queensland human settlements: Issues and context', unpublished report for the South East Queensland Climate Adaptation Research Initiative, Griffith University, Queensland.

McKenzie Hedger, M, Brown, I, Connell, R and Gawith, M 2000, *Climate Change: Assessing the Impacts – Identifying Responses. Highlights of the First Three Years of the UK Climate Impacts Programme,* Department of Environment, Transport and the Regions, London, viewed 31 March 2017, https://www.google.com.au/url?sa=t&rct=j&q=&esr

c=s&source=web&cd=1&ved=0ahUKEwiq0a_RkIDTAhVL3WMKHcWfDFQQFgg
ZMAA&url=https%3A%2F%2Fora.ox.ac.uk%2Fobjects%2Fuuid%3Ae9832ef0-b133-
456a-a48c-172aa5e94b1d%2Fdatastreams%2FATTACHMENT01&usg=AFQjCNEv3
nSUQnrGBMG0GyyYj1PWoBWJvA&sig2=aYgzBvTApTCaYvI0PT6MHg&bvm=b
v.151426398,d.cGc&cad=rja./

McLean, I 2008, 'Climate change and UK politics: From Brynle Williams to Sir Nicholas
Stern', *The Political Quarterly*, vol. 79, no. 2, pp. 184–193.

Maher, S 2011 'Climate change department for chop: Hockey', *The Australian*, 4 August,
viewed 5 August 2013, www.theaustralian.com.au/national-affairs/climate/climate-
change-department-for-chop-hockey/story-e6frg6xf-1226108027421/.

Marris, E 2007, 'Australia warms to climate change', *Nature Reports–Climate Change*,
November, vol. 6, pp. 90.

Mercer, D, Christesen, L and Buston, M 2007, 'Squandering the future–Climate change,
policy failure and the water crisis in Australia', *Futures*, vol. 39, nos. 2–3, pp. 272–287.

Moore, C 2013a, 'To win the battle for the consumer, Cameron must cut taxes soon', *The
Daily Telegraph*, 2 September, viewed 21 July 2014, www.telegraph.co.uk/news/
politics/conservative/10339560/To-win-the-battle-for-the-consumer-Cameron-must-
cut-taxes-soon.html/.

Nelson, DR, Adger, WN and Brown, K 2007, 'Adaptation to environmental change:
Contributions of a resilience framework', *Annual Review of Environment and Resources*,
vol. 32, pp. 395–419.

National Emergency Management Committee [NEMC] 2010, *National Emergency Risk
Assessment Guidelines*, Attorney General's Department, Canberra.

Norman, B 2012, 'World Bank calls for greater climate preparedness – in Australia, planning
unravels', *The Conversation*, 20 November, viewed online 31 March 2017, http://
theconversation.edu.au/world-bank-calls-for-greater-climate-preparedness-
inaustralia-planning-unravels-10807./

Oberthur, S and Ott, HE 1999, *The Kyoto Protocol: International Climate Policy for the 21st
Century*, Springer, Berlin.

OFWAT 2010, *Adaptation to Climate Change: Statutory Reporting Information Note*, The Water
Services Regulation Authority, August 2010, viewed 30 October 2013, http://www.
ofwat.gov.uk/sustainability/climatechange/prs_inf_20100820climate./

OFWAT 2014b, 'History of the water and sewerage sectors', website, viewed 8 December
2014, www.ofwat.gov.uk/industryoverview/history/.

Organisation for Economic Co-operation and Development [OECD] 1999, *National
Climate Policies and the Kyoto Protocol*, OECD, viewed 16 October 2013, www.oecd-
ilibrary.org.libraryproxy.griffith.edu.au/docserver/download/9799101e.pdf?expires=13
81820226&id=id&accname=ocid53013929&checksum=C507522A6EF41AD6929E7
EFD74957222/.

Peters, EJ and McEntire, DA 2014, *Emergency Management in Australia: An Innovative,
Progressive and Committed Sector*, Federal Emergency Management Agency website,
viewed 2 January 2017, https://search.usa.gov/search?affiliate=netc&query=australia&o
p=Search/.

Pielke Jr, RA 2009, 'The British Climate Change Act: A critical evaluation and proposed
alternative approach', *Environmental Research Letters*, vol. 4, no. 2, pp. 1–7.

Porritt, J 1989, 'The United Kingdom: The dirty man of Europe?', *RSA Journal*, vol. 137,
no. 5396, pp. 488–500.

Preston, BL, Rickards L, Dessai S and Meyer, R 2013, 'Water, Seas, and Wine: Science for
Successful Climate Adaptation', in Moser, SC and Boykoff, MT (eds), *Successful*

Adaptation to Climate Change: Linking Science and Policy in a Rapidly Changing World, Routledge, Abingdon.

Preston, BL, Westaway, RM and Yuen, EJ 2011, 'Climate adaptation planning in practice: An evaluation of adaptation plans from three developed nations', *Mitigation and Adaptation Strategies for Global Change*, vol. 16, no. 4, pp. 407–438.

Productivity Commission 2010, *Strengthening Evidence Based Policy in the Australian Federation, Volume 1: Proceedings*, Roundtable Proceedings, Productivity Commission, Canberra.

Pusey, M 1991, *Economic Rationalism in Canberra: A Nation-Building State Changes Its Mind*, Cambridge University Press, Cambridge.

Queensland Floods Commission of Inquiry [QFCI] 2011, *Queensland Floods Commission of Inquiry: Interim Report*, QFCI, Brisbane, viewed 3 April 2013, www.floodcommission. qld.gov.au/publications/interimreport/.

Queensland Floods Commission of Inquiry [QFCI] 2012, *Queensland Floods Commission of Inquiry: Final Report*, QFCI, Brisbane, viewed 3 April 2012, www.floodcommission.qld. gov.au/__data/assets/pdf_file/0007/11698/QFCI-Final-Report-March-2012.pdf.

Queensland Government 2009b, *ClimateQ: Toward a Greener Queensland*, Department of Environment and Resource Management, Brisbane, viewed 13 August 2013, http://rti. cabinet.qld.gov.au/documents/2009/may/climateq%20toward%20a%20greener%20 qld/Attachments/ClimateQ_Report_web_FINAL_20090715.pdf.

Queensland Government 2012a, *Queensland Coastal Plan*, Department of Environment and Heritage Protection, Brisbane, viewed 13 August 2013, www.ehp.qld.gov.au/ coastalplan/pdf/qcp-web.pdf.

Queensland Government 2014b, *Queensland Strategy for Disaster Resilience*, Queensland Government website, viewed 2 January 2016, http://dilgp.qld.gov.au/local-government/community/community-recovery-and-resilience.html/.

Queensland Government 2016a, *Population Growth Highlights and Trends: Queensland, 2016 Edition*, Queensland Government Statistician's Office, Queensland Treasury, Brisbane, viewed 12 November 2016, www.qgso.qld.gov.au/products/reports/pop-growth-highlights-trends-qld/index.php/.

Queensland Government 2016b, *Adapting to Climate Change*, Queensland Government website, viewed 20 December 2016, www.qld.gov.au/environment/climate/adapting/.

Queensland Government 2016c, *Disaster Management: Policies, Guidelines and Forms*, Queensland Government website, viewed 2 January 2017, www.disaster.qld.gov.au/ Disaster-Resources/pages/PGF.aspx/.

Queensland Government 2016d, *RACQ Get Ready Queensland*, Queensland Government website, viewed 2 January 2017, https://getready.qld.gov.au/.

Reser, JP, Bradley, GL, Glendon, AI, Ellul, MC and Callaghan, R 2012, *Public Risk Perceptions, Understandings, and Responses to Climate Change and Natural Disasters in Australia, 2010 and 2011*, National Climate Change Adaptation Research Facility, Gold Coast, Queensland, viewed 17 March 2015, www.nccarf.edu.au/publications/ public-risk-perceptions-second-survey/.

Richards, D and Smith, MJ 2002, *Governance and Public Policy in the UK*, Oxford University Press, Oxford.

Rootes, C 2008, 'The first climate change election? The Australian general election of 24 November 2007', *Environmental Politics*, vol. 17, no. 3, pp. 473–480.

Ross, A and Dovers, S 2008, 'Making the harder yards: Environmental policy integration in Australia', *The Australian Journal of Public Administration*, vol. 67, no. 3, pp. 245–260.

Rudd, K 2013, '1st Leaders Debate for the 2013 Federal Election', opening speech broadcast on ABC, 11 August, viewed 13 August 2013, www.abc.net.au/iview/#/series/12771/.

Rydin, Y 2003, *Urban and Environmental Planning in the UK*, 2nd edn, Palgrave Macmillan, London.

Sharp, R 1998, 'Responding to Europeanisation: A Governmental Perspective', in Lowe, P, and Ward, S (eds), 1998, *British Environmental Policy and Europe: Politics and Policy in Transition,* Routledge, London.

Sheail, J 2002, *An Environmental History of Twentieth-Century Britain*, Palgrave, New York.

Solesbury, W 2001, '*Evidence Based Policy: Whence it Came and Where it's Going – Working Paper 1*', ESCRC UK Centre for Evidence Based Policy and Practice, Queen Mary, University of London.

Spearritt, P 2009, 'The 200km city: Brisbane, the Gold Coast and the Sunshine Coast', *Australian Economic History Review*, vol. 49, no. 1, pp. 87–106.

Talberg, A, Hui, S and Loynes, K 2013, *Australian Climate Change Policy: A Chronology*, Australian Department of Parliamentary Services, 2 December 2013, viewed 3 February 2015, http://parlinfo.aph.gov.au/parlInfo/download/library/prspub/2875065/upload_binary/2875065.pdf;fileType=application/pdf.

Tangney, P 2015, 'Brisbane City Council's Q100 Assessment: How climate risk management becomes scientised', *International Journal of Disaster Risk Reduction*, vol. 14, no. 4, pp. 496–503.

Tangney, P and Howes, M 2016, 'The politics of evidence-based policy: A comparative analysis of climate adaptation in Australia and the UK', *Environment and Planning C: Government and Policy*, vol. 34, no. 6, pp. 1115–1134.

Taylor-Gooby, P 2012, 'Root and branch restructuring to achieve major cuts: The social policy programme of the 2010 UK coalition government', *Social Policy and Administration*, vol. 46, no. 1, pp. 61–82.

Thompson, E 1980, 'The 'Washminster' Mutation', in Weller, P and Jaensch, D (eds), *Responsible Government in Australia*, Drummond & the Australasian Political Studies Association, Victoria.

Tompkins, EL, Adger, WN, Boyd, E, Nicholson-Cole, S, Weatherhead, K and Arnell, N 2010, 'Observed adaptation to climate change: UK evidence of transition to a well-adapting society', *Global Environmental Change*, vol. 20, pp. 627–635.

Toynbee, P 2007, 'Nimbys can't be allowed to put a block on wind farms', *The Guardian*, 5 January, viewed 24 October 2013, www.theguardian.com/commentisfree/2007/jan/05/comment.politics/.

UK Government 2014, *Resilience in Society: Infrastructure, Communities and Business*, Cabinet Office, Gov.uk, 10 December 2014, viewed 16 January 2017, www.gov.uk/guidance/resilience-in-society-infrastructure-communities-and-businesses/.

Victorian Bushfires Royal Commission [VBRC] 2010, *Final Report: Summary*, Parliament of Victoria, Melbourne.

Volunteering Queensland 2015, *Queensland Disaster Management Arrangements*, Volunteering Queensland website, viewed 2 January 2017, www.emergencyvolunteering.com.au/qld/disasterready/qdma/.

Walker, KJ 2002, 'Environmental Policy in Australia', in Desai, U (ed.), *Environmental Politics and Policy in Industrialized Countries*, MIT Press, Cambridge, MA.

Walker, KJ 2012, 'Australia's Construction of Environmental Policy', in Crowley, K and Walker, KJ (eds), *Environmental Policy Failure: The Australian Story*, Tilde University Press, Victoria, Australia.

Wanna, J and Weller, P 2003, 'Traditions of Australian governance', *Public Administration*, vol. 81, no. 1, pp. 63–94.

4 Climate adaptation evidence for policy

Because scientists do not recognize a legitimate role for values in science (it would damage "objectivity"), scientists avoid discussion of the choices they make.

Heather Douglas

In this chapter I explain the epistemological character of adaptation evidence and adaptation policy problems, to provide the reader with a framework for understanding climate-related knowledge and expertise. This understanding will be helpful for considering the tensions between expert and political authority for climate adaptation, as described in Chapter 5. It will also provide a point of comparison for policy players' perceptions of climate science and policy evidence to be described in Chapter 6, and the linear-technocratic expectations of policymaking that contribute to processes of politicisation and scientisation in England and Queensland, described in Chapter 7.

As Hoppe (2005) and Owens et al. (2004) amongst others argue, knowledge utilisation studies suggest that policy evidence is often used selectively or tactically to legitimise an extant political position and any expert findings that favour an alternative position are often lost or dropped from the evidence-based narrative or argument. This tendency is particularly relevant for areas of policy development such as climate adaptation that are highly uncertain and complex and for which experts often cannot provide definitive knowledge. If, as I argue, governments pursue the evidence-based mandate only to the extent that their evidence aligns with prevailing norms and politics, then it would seem that the legitimacy of evidence and the legitimacy of policy issues themselves are interconnected in important ways.

The difficulties of establishing and maintaining legitimacy for ancillary policy issues is an ongoing challenge for policy advocates, particularly in relation to environmental issues that have rarely topped the political agendas of Australia or the UK (Jordan, 2002b; Walker, 2002). Issues like climate adaptation must compete for attention with a range of other social and environmental problems that do not preoccupy the minds of the political executive with the same frequency or intensity as, say, the economy, education, crime or health services. In Chapter 3, I described how UK and Queensland government's adherence to the evidence-based mandate has been conditional upon coherence between these governments' political priorities and available expert knowledge. And while politically acceptable evidence may not be a *sufficient* precondition for policy action, it is nonetheless often a *necessary* one. Maintaining legitimacy for an ancillary issue like climate adaptation relates not just to its palatability in normative or political terms, but also upon the ability of corroborating evidence to demonstrate that the issue is worthy of scarce political resources. As one UK policy player suggested in the course of researching this book:

> [If] it's too uncertain an issue to deal with, you only have to drop out of the top ten priorities to really not be doing very much at all.
>
> (UK-Policy Scientist 7)

This sentiment, echoed by a number of others, highlights the difficulties of maintaining policy legitimacy for an issue like climate change adaptation in the absence of legitimate, credible and salient evidence (UK-Policy Scientists 6, 7, 8; UK-Policy Players 2, 3, 7).

In this chapter, therefore, my explanations of the epistemology of the available science, as well as the character of policy problems themselves, are an attempt to explain the origins of some of the difficulties encountered when balancing the competing requirements for technically robust, instrumentally usable and politically acceptable knowledge.

The development of climate change science and policy evidence

In Chapter 2, I described the importance of the work of constructivist scholars for understanding the role of evidence and expertise in policy decision-making. In this chapter I use these constructivist arguments again to characterise the information available for adaptation policymaking. It is worth noting that this characterisation is currently more relevant to the UK case-studies in this book since, as argued in Chapter 3, in England the development of political and expert consensus under the Climate Change Act (2008) has effectively legitimised climate change science for use in policymaking. As a result of this consensus, policy evidence develops based on the same pool of available climate change science. In the Queensland case, by contrast, this characterisation is less applicable at the current time due to the lack of legitimacy for climate *change* policy, and by association, climate change science. As I discuss further in Chapter 7, the process by which climate adaptation becomes scientised in Australia relates more to the selective use of alternative sets of credible evidence rather than varying interpretations of the same evidence base, as in the UK.

As suggested by the work of Hertin et al. (2009), Nilsson et al. (2008) and Owens et al. (2004), the knowledge used to rationalise and legitimise decisions is often an interpretation of data that is open at various points to normative and even political influence. Indeed, within many branches of evidence-based policymaking, the evidence that is used – derived through *ex ante* appraisal techniques such as cost-benefit analysis and regulatory impact assessment – may have limited direct input from technical experts and are conducted by bureaucrats for policymakers (see, for example, Turnpenny et al. 2009; Radaelli and Meuwese, 2010). This is generally not the case for climate change. Perhaps due to its complex and contentious nature, scientific experts have been more directly involved in the derivation of policy evidence, and for the examples used here, considerable emphasis has been placed on their technical and scientific rigour.

Nonetheless, political influence upon this evidence can come from underlying subjective choices required in the development of climate science (Kellow, 2009); in the focus and development of subsequent adaptation science (Preston et al., 2013), as well as from the interpretive nature of policy evidence development itself (Nilsson et al., 2008; Owens et al., 2004). Yet, under a linear-technocratic policymaking schema this final output is often construed as impartial and independent expert evidence (see for example, DEFRA, 2012a: p. 3, 7; NEMC, 2010: p. 4). Since climate and adaptation science is necessarily uncertain about the future and is often also uncertain about the precise nature of local climate impacts, and because scientific messages are so dependent on how uncertainty is understood and presented, climate adaptation policy is particularly dependent on interpretive or contingent understandings that no level of expertise can render impartial. The necessity for value judgements and, therefore, the propensity for political influence when deriving policy evidence is important to understand why and how, as I argue here, it may be perceived as lacking credibility, legitimacy and/or salience for the purposes of adaptation policymaking (see also Chapter 6).

Here I describe a four-part process by which evidence develops for climate change adaptation. This conceptual framework explains important distinctions between climate science, adaptation science and policy evidence, the types of value judgements prevalent in the development of expert knowledge about climate change and, therefore, the extent to which each of these types of climate change evidence can be considered objective knowledge.

What is climate science?

The field of climate science has a long and chequered history. As Edwards (2010) and Miller and Edwards (2001) have described in much detail, climate science has been developing since the nineteenth century and was, for most of this time, a practice of record keeping and statistical analysis from a rather disparate and disjointed global community of both professional scientists and hobbyists. Climatology (what subsequently became known as climate science) has always found patronage from governments, often for military purposes. However, a number of factors have ensured that climatology has always had difficulties maintaining statistically compatible and geographically consistent records across the globe: global politics, two world wars, uneven socio-economic development and the evolution of concurrent science and technologies for transport and telecommunications. As such what we know about past climate, and the influence of anthropogenic forcing upon it, has been built from a patchwork of data that has required innovative means of statistical integration to provide global and regional coherence. As Edwards (2010) describes:

> We have not one data image of the global climate, but many. The past, or rather what we can know about the past, changes. And it will keep right on changing … Global data images have proliferated, yet they have also converged. They shimmer around a central line, a trend that tells us that Earth has already warmed by about 0.75C … since 1900.
>
> (Edwards, 2010: p. xiii)

It was not until the 1970s or so that climate science was transformed through the use of computers. The advent of computer technology allowed data to be used and developed through modelling and parameterisation to create global knowledge in ways that revolutionised how the concept of climate was interpreted and understood. Although the precise details of how these models and modelling techniques developed are beyond the scope of this book (see instead, Petersen, 2012; Edwards, 2010; Miller and Edwards, 2001), it is worth noting that the term 'modelling' refers to a number of different practices. Modelling involves the interpretation and manipulation of readings from weather instruments, the simulation of climate systems, the simulation of Earth's bio-geophysical systems more generally, and the integration of weather forecasting methods, climate simulations and socio–economic data in ways that allow climate scientists to present internally consistent simulations of future climate.

Computer modelling has become the 'virtual laboratory' of climate science through which hypotheses of anthropogenic climate forcing can be tested (Kellow, 2009). However, when it comes to anticipating future climatic change, these models can only be wholly verified or validated in a *post hoc* way. Due to the highly complex (and potentially chaotic) nature of Earth's planetary systems we only truly know that climate models are accurate in relation to their ability to simulate *observed* climate over meaningful timescales (Frigg et al., 2013a; Oreskes et al., 1994). What the climate science community produces for policy decision-makers, therefore, consists of conclusive evidence of past anthropogenic influence on the Earth's climate, alongside necessarily uncertain projections that present potential future climate scenarios. These climate projections are produced for various points in the future under alternative sets of assumptions concerning the emission of GHGs and the climatic, bio-geographical and social-ecological characteristics and dynamics of Earth's natural regulatory systems (Edwards, 2010). Climate change scenarios and projections are developed using what are known as General Circulation Models (GCMs) and more recently, through the advent of more complex Earth System

Models (ESMs) that create simulations of global climate and potential future climate change.

Here, I describe GCMs, ESMs and experts' use of them to project future climate change as 'trans-scientific' (Weinberg, 1972). That is, climate scientists ask what appear to be purely scientific questions through the use of these models, which inevitably require more than scientific reasoning to answer. Weinberg identified three indicative (though probably not exhaustive) ways in which scientists' policy advice may be trans-scientific. In terms of climate-related science, I suggest that these can manifest in the following ways:

1   Scientists use epistemic shortcuts requiring subjective value judgements when developing data models or when drawing conclusions about cause and effect relationships between variables. Such shortcuts are necessary since, to substantiate empirical results with epistemologically adequate confidence levels would otherwise require unrealistically time-consuming or impractical levels of testing and observation.
2   Scientists make subjective value judgements in the development of proxy data and algorithmic approximations of complex interactions because there is too much variability, or there are too many variables at play in a system to allow for direct determination of system dynamics and end-points based on simple cause and effect relationships.
3   Scientists may make significant value judgements in the course of presenting their science because, in order to provide useful advice, social or ethical value choices are required to make comparisons that allow for meaningful scientific answers.

Climate change science displays each of these characteristics.

In relation to the first of these trans-scientific modes, for example, the processes of raw data-modelling associated with the aggregation and organisation of a global, heterogeneous and incomplete collection of weather and climate records requires trans-scientific judgements to ensure this data is coherent and usable for the purposes of constructing climate simulations. These judgements relate to how that raw data should be organised and how to account for the (sometimes considerable) gaps in available information in a way that is consistent with the body of evidence as a whole.

In relation to the second mode, similarly, climate models must account for such complexity in their application of physical theory during the construction of models that semi-empirical parameterisations of key variables or sub-systems are required to allow models to approximate climate outcomes. This parameterisation contributes to what Edwards (2010: p. 281) refers to as the *reproductionist* mode of climate science:

> Reproductionism seeks to simulate a phenomenon, regardless of scale, using whatever combination of theory, data, and 'semi-empirical' parameters may be required.

Modes one and two of trans-scientific activity will often involve epistemic and cognitive value judgements (Douglas, 2009) (see Chapter 2) that are subject to the standards and methodological norms of good scientific practice. However, as discussed below, it is also possible that these trans-scientific judgements may require, or be influenced by, non-epistemic value choices.

Third, and perhaps most significantly, in order to make reasonable statements about anthropogenic climate *change*, climate scientists must make normative judgements about what models' estimated change is relative to, what factors that change depends upon, and what the significance of that change is. What is normal climate? How will the global climate change over the coming century? At what point may climate change become dangerous? These questions are asked by, or of, the climate science community and are engaged at the point of presenting and drawing conclusions about climate simulations. They require not just a range of subjective assumptions and assertions requiring epistemic value judgements (like those described for trans-science modes one and two above) as a result of the many uncertainties about the complex ways in which bio-geophysical and social-ecological systems interact (Petersen, 2012; Edwards, 2010), they also require or may otherwise be influenced by non-epistemic judgements regarding how things are (or ought to be) and how we envision the future (Hulme, 2009).

It is worth elaborating on the concept of a climatic baseline to understand the extent of models' trans-scientific character in this regard. Climate change models are designed to inform our understanding of change relative to some baseline of what is considered to be 'normal' climate. In the case of IPCC modelling, normal climate is currently based on statistical assessment of a 30-year time period between 1986 and 2005 (Stocker et al., 2013: p. 79). However, climate is both a statistical and a cultural construct. Edwards (2010: p. xiv) describes climate as 'the history of weather', but as suggested by Hulme et al. (2009), Strauss and Orlove (2003), Fagan (2000) and Glacken (1990) amongst others, climate is much more than simply an aggregation of weather statistics as agreed by the World Meteorological Organisation (Edwards, 2010).

Although climate can be defined scientifically as the accumulation of weather statistics over a 30-year period, it is also a concept that pervades our natural histories of the world, the stories we tell, and the lives that we lead. Climate is a central but often unstated constituent of the many cultural traditions of human societies. On a planet where the climate is continually changing, interpretations of 'normality' (i.e. which 30-year period of weather we consider to be *normal*), therefore, are ultimately based on opinions about how the climate ought to be, which the laity could justifiably claim to have equally valid input to; concepts of normality pervade everything we do and know, and not just the realms of science. As discussed in Chapter 2, any unqualified assumption that science should dictate what *should be* 'normal' climate, is tantamount to the scientific community annexing an entirely non-epistemic judgement about what we as a global community value. Making any such assumptions is also to imply, erroneously, that science is immune from socio-political influence and therefore the only source of valid knowledge about the world. Hopefully, climate scientists – even those of a more positivist persuasion – will concede that such *scientistic* sentiments do not hold water.

Model input such as the assertion of climate normalcy can have a significant bearing on both experts' and policymakers' subsequent interpretations about the range and severity of potential future climate *change* impacts, depending on how this construct is formulated and used (Hulme et al., 2009). The idea of climate change relative to what is 'normal' is the prime example, therefore, of how, although climate science may be presented as wholly objective, and is primarily driven by scientific methods of data analysis and modelling, some of the methods used by climate scientists also subsume social and ethical value choices.

Trans-scientific questions concerning climate change are determined in part by processes of data-modelling and parameterisation that cannot be answered solely through observation (in advance of climate change impacts actually occurring), by testing and falsifying hypotheses, or by actual repeatable experiment other than through the development and interpretation of simulation models, which themselves cannot be wholly verified or validated in advance of future climate change (Frigg et al., 2013a; Oreskes et al., 1994). Climate change modelling also requires, therefore, subjective, normative judgements in the derivation, selection and interpretation of data that either directly require (e.g., in the choice of 'normal' climate), or can be influenced by (e.g., in the choice, interpretation and manipulation of datasets, or the parameterisation of climate variables), socio-cultural as well as technical reasoning (Fischer, 2009). Trans-scientific judgements, I argue, are no more immune from the socio-cultural ideals of scientists than scientists are immune from confirmation bias (Ioannidis, 2005). As a result, trans-scientific choices during evidence development can transcend what we would traditionally consider to be acceptable epistemic or cognitive value judgements (see Chapter 2) (Douglas, 2009).

A word of caution is necessary at this point. It is not my intention here to exaggerate the normative components of climate science, nor to use them to dismiss climate change models as somehow constituting politics by another name. The partiality of trans-science does not diminish the usefulness of climate models per se, nor necessarily call into question their authority or legitimacy as expert evidence for informing public decisions, *as long as they are used appropriately*. The accuracy of GCMs/ESMs in recreating past climate and for understanding the influence of GHGs to date has been largely validated, irrespective of what climate baseline is used to understand potential future changes (Stocker et al., 2013).[1] On this basis we have good reason to be alarmed by climate change and to pay heed to projections of future climate hazards. Indeed without many of the necessary value judgements made during the choice and interpretation of meteorological data and in the construction of weather and climate models, we would not know very much at all about the global climate or the phenomenon of anthropogenic warming that has been empirically validated by other means (e.g., through the use of satellite data). Nonetheless, the trans-scientific nature of climate models does highlight the largely intractable uncertainties associated with climate change modelling, such that they cannot provide unequivocal facts about *future* climate (Frigg et al., 2013a; Edwards, 2010).

The social and ethical (non-epistemic) value choices involved at every stage in the development and presentation of climate change evidence inevitably point to the possibility of a politicisation-by-process of climate science for policy. This is *not* to suggest that such politicisation is necessarily sinister, but that because climate scientists are required to make non-epistemic value judgements, it follows logically that these judgements could inevitably be unduly coloured by normative/political viewpoints. Climate scientists are as prone to cognitive biases such as confirmation bias[2] as any other branch of the scientific academy (Ioannidis, 2005; Nickerson, 1998), and perhaps more so, given the hyper-politicised nature of the climate change problem. Indeed, it appears to have become positively taboo within the broader climate change academy to question the central hypotheses associated with anthropogenic warming; to question the consensus interpretations of the available climate science; or even, to accurately portray the available consensus science where it does not align with climate change advocates warnings about climate change disaster, as established scholars such as Roger Pielke Jr and Judith Curry have learned to their cost (Pielke Jr, 2015; Waldman, 2017).

It is also worth noting that the politicisation of climate science outputs has been investigated by both scientific and social-scientific scholars in recent years as a more sinister development in the interplay between politics and science. Judith Curry (2013), for instance, highlights trans-scientific characteristics of climate science when she criticises the lack of transparency in the subjective/normative judgements made by the IPCC about the validity of different datasets when accounting for global temperature changes over recent decades. Likewise, Kellow (2009) presents a detailed case implicating US climate scientists in the doctoring of data models, he claims, as a means of promulgating the anthropogenic warming theory due to normative/political rather than scientific motives. I won't dwell on these arguments here because they are not centrally relevant to the arguments in this book, and I believe that their significance is over-stated by those who deny the validity of the anthropogenic warming theory. However, beyond any specific cases, I believe it is worth considering that political influences *can* be prevalent in the derivation of climate science, and that the normative components of climate science are either not recognised or not acknowledged by the scientific community.

Although this is undoubtedly a controversial line of questioning, it should not be. To give a fair account of the role of values in climate change evidence for policy more broadly, we must recognise the potential partiality of this knowledge, while at the same time recognising that all empirically valid science includes value judgements of some sort or other (Douglas, 2009). While there is clearly a political expediency in moving beyond the arguments of those who entirely deny anthropogenic climate change, in order to effectively address the associated hazards, this should not mean that scientists and other scholars within the academic community should be entirely discouraged from probing established hypotheses and conjectures or questioning the normative components of climate change science. That some scholars appear to have been suppressed in these activities, suggests that political influences have infiltrated the scientific community in scientifically unhealthy ways.

★ ★ ★ ★ ★

The concept of trans-science is useful because it highlights the limits of positivist or reductionist concepts for understanding the nature of climate science and climate change. Trans-science also demonstrates, however, the inadequacy of scientific expertise on its own, when seeking to understand pluralistic and location-specific adaptation problems and, therefore, for informing adaptation policy. As I discuss in the second part of this chapter, even with hypothetical (though practically impossible) advancements that could ensure their near-absolute accuracy (Frigg et al., 2013a, 2015), climate change projections cannot resolve the types of contingent and contextual questions posed in the development of evidence for adaptation policy. Suffice it to say at this point, however, that the scenarios produced using GCMs/ESMs are necessarily of limited use for decision-making on their own.

As described in Chapter 6, policy players are only likely to find a global projection of future climate useful and usable in the sense that they can begin to understand that a problem exists, and at that, in a rather abstract way, since trends of variables such as global or regional average temperature changes are difficult to conceptualise and to understand at meaningful levels and scales of governance (Wilbanks and Kates, 1999). So it is that climate scientists must begin to consider the needs of decision-makers in order to make this information more usable (i.e., salient, if not also legitimate) to the broadest range of users.

A first step in making more usable science from GCMs/ESMs is a process known as 'downscaling' whereby model outputs are manipulated to provide climate change projections at finer resolutions and for smaller geographical areas. Such regional projections are potentially more useful and usable than the first outputs of GCMs/ESMs since they provide information at geographic scales that are closer to those at which government decisions are made. However, as we shall see in Chapter 6, they may still provide rather abstract information for many users trying to understand precisely what climate change has in store for them and what that means for policymaking. Moreover, there remain significant questions about the reliability of downscaling techniques for producing decision-relevant predictions of future climate (Frigg et al., 2013b). Nonetheless, what may ensue at this point is a process of *post-normal* scientific development (Hulme and Dessai, 2008; Funtowicz and Ravetz, 1993) whereby climate science outputs are co-designed between climate scientists and other expert (and potentially also non-expert) users of this science, to ensure that it provides the types of information that can be most meaningful or salient for understanding climate impacts and making decisions at relevant geographic, jurisdictional and temporal scales (Mullan et al., 2013; Cash et al., 2003).

As argued by Cash et al. (2003), salience is an important characteristic of effective knowledge for decision-making. Delivering salient evidence often requires one or both sides of the knowledge-action (or science-policy) interface to bridge the divide between these two realms, in order for experts to effectively understand the needs of knowledge users (see Chapter 5). For instance, the UK Climate Projections 2009 (UKCP09) used downscaling techniques to provide projections at regional and even local scales (25 km^2 grid squares) and in a probabilistic format that would provide decision-makers with likelihoods of change of a range of climate variables. These projections, derived from an ensemble of GCMs, were developed through a post-normal scientific approach (see Chapter 2) that sought the views of potential users to understand the optimal format and content of scientific outputs to make them most usable (Porter and Dessai, 2016; Hulme and Dessai, 2008).

The epistemological distinction between the trans-science involved in the development of GCMs/ESMs and the post-normal science of downscaled climate change scenarios and projections may be best characterised in terms of the contribution of a core-set of experts versus other experts and non-experts in their development. Whereas the subjective/normative decisions involved in trans-science are (for the most part)[3] taken by the various climate-related experts who contribute to the development of these complex models, the subsequent post-normal science involved in developing scenarios and projections should usefully involve consultation and co-design between climate experts, other scientific and technical experts, as well as non-expert users of this information.

Nonetheless, this post-normal science often requires even further interpretation in order to provide meaningful knowledge to guide policy decisions. As described in Chapter 6, testimony from policy players in the UK and Australia suggests that this information, of itself, still lacks salience and legitimacy and is rather limited in the ways in which it can directly inform policy decision-making. Post-normal climate science in the form of climate projections – such as developed for UKCP09 – are often too complex, too abstract and/or simply too uncertain for policy players to use as a means to instrumentally direct policy on their own. At this point it must be interpreted through the development of both *adaptation science* and *policy evidence*.

What is policy evidence?

For the purposes of this research, I define policy evidence as a subset of the broader body of technical evidence available for decision-making (see Chapter 2). Post-normal climate change science is often (though not necessarily) interpreted by experts in the development of adaptation science.[4] These scientific outputs are then interpreted through further coproduction processes between experts and non-expert policy players for the development of policy evidence. In this latter effort, climate science and adaptation science may be combined with other scientific and socio-economic data sets of environmental quality indicators, standards or thresholds, alongside policy players' judgements about how to understand climate hazards and potential impacts, in order to derive meaningful and politically acceptable policy evidence.

Policy evidence, therefore, is the form in which science and other expert knowledge is presented to government decision-makers to allow them to understand a problem like climate change in terms that they can relate to and understand. What distinguishes policy evidence from other forms of decision-making knowledge is that it is overseen and/or commissioned by government for the purposes of informing policy and is usually coproduced between experts and non-expert policy players for the purposes of making technical evidence politically acceptable.

As an example, Preston et al. (2013) describe how the UK water industry and the Environment Agency of England and Wales (a non-departmental government regulatory body) have used downscaled climate projections (post-normal climate science) in conjunction with hydrological data to forecast how river flows might be affected by a changing climate at various points throughout the twenty-first century, in order to inform water resource management plans. This constitutes what Moss et al. (2013) and Preston et al. (2013) refer to as *adaptation science* and is often undertaken by the research community in tandem with climate science outputs. While adaptation science may have trans-scientific constituents, it is also directed and subject to consultation from both experts and non-experts concerning research focus, the format of scientific outputs and where investments in science should be allocated (Preston et al., 2013; Meyer, 2011). It can, therefore, constitute post-normal science. However, adaptation science such as this may also be a core constituent of *policy evidence* alongside climate science, and may be overseen by government in order to inform the design and implementation of policy. In the case presented by Preston et al. (2013) above, the design of water resource management plans was overseen by a government institution as part of the regulatory cycle of the water industry. Adaptation science such as this may, therefore, fall into the category of either post-normal science or coproduced policy evidence. The cases described in this book suggest that adaptation science outputs are often used by policy players either directly or indirectly (i.e., through further interpretation using methods such as risk assessment) in the development of policy evidence.

Some policy evidence, however, is less obviously scientific in its character and requires considerably more obvious reliance on bureaucratic deliberation and political interpretation. Variously described as 'policy appraisal' (Owens et al., 2004), 'policy assessment' (Hertin et al., 2009) or simply policy 'evidence' (DEFRA, 2012b; NEMC, 2010), this type of knowledge involves the quantitative or semi-quantitative assessment of risk, costs and benefits of a changing climate on government's objectives and priorities. This type of policy evidence may use the outputs of climate and adaptation science as well as processes of deliberation between experts and non-experts, and is usually the final hurdle allowing complex expert evidence to be used for policymaking. That is to say, this type of policy evidence requires such a degree of subjective and normative development and interpretation, in the imposition of limits of scale (jurisdictional, temporal and/or geographical), in both the choice and use of science and other expert research, arguably much more than its trans- and post-normal scientific antecedents, that can have significant political implications. Owens et al. (2004, p. 1943) appear to concur with the idea of policy evidence as partially political in its derivation when they argue that:

> an important role for [policy] appraisal (by design or by default) may be that of providing spaces for dialogue and learning in the making of policies and decisions.

Indeed as I describe for the case of the UK's Climate Change Risk Assessment (UKCCRA) (DEFRA, 2012a,b) and the derivation of Q100 flood metrics in Queensland in Chapter 7, the means by which this evidence is derived often constitutes a process of 'coproduction' between expert and non-expert participants (Jasanoff and Wynne, 1998).

Policy evidence takes a number of different forms, but in most cases its development involves a prescribed process of data collection, interpretation and deliberative appraisal to enable policymakers to understand important considerations when making a decision (Owens et al., 2004). Policy evidence is often produced through the assessment of risks and opportunities from a potential hazard, or to understand the costs and benefits of a proposed action to address them. Importantly, such evidence has, to date, been developed and used within a linear-technocratic schema of policymaking so that these evidence outputs are construed (intentionally or otherwise) as objective impartial research and advice to government (see for example, NEMC, 2010: p. 4). These scientific and economic analyses, however, require not simply authoritative expert input (which is itself often dependent on both epistemic and non-epistemic value judgements (Douglas, 2007)), but also a broader socio-cultural reasoning usually involving non-expert policy players to understand the implications of potential hazards upon society, economies and ecologies. Under the linear-technocratic schema, this suggests that:

1   non-experts are involved in the development of what is ultimately interpreted as objective authoritative advice; and/or,
2   experts may be extending their authority (inadvertently or otherwise) beyond their legitimate field of expertise (i.e., that which they have *privileged* knowledge and expertise about).

Policy evidence development, as described in the literature (Hertin et al., 2009; Nilsson et al., 2008; Owens et al., 2004; Clark and Majone, 1985) and further in later chapters here, speaks to the inherently political potential of the knowledge used for climate adaptation policy. This evidence is subject to both a politicisation-by-process (as a result of an inevitable normative input to expert climate and adaptation science required to make it coherent and usable) and also potentially a politicisation-by-agency (a deliberate adjustment of evidence outputs to align with political priorities). These political influences are then also shaped by established methodological norms and the deliberative nature of policy evidence development, as well as by the contingent nature of adaptation problems themselves. In short, policy evidence involves the development of consensus between available science and assessors' norms and implicit values about the world, as well as potentially involving their political priorities.

As I describe in greater detail in later chapters, policy evidence development is a process of negotiation and coproduction between a range of expert and non-expert policy players about what the climate problem is, who it affects and how, and what can be done to resolve it. Its derivation requires inevitable socio-cultural, non-epistemic value judgements in the assessment of impacts, risks, costs and benefits concerning the vulnerability, adaptation and resilience 'of what and for whom'. Importantly, given its deliberative character, policy evidence does not necessarily correlate with the views of the individual experts contracted to develop it, yet relies upon their expert authority for its legitimacy. This characterisation suggests, therefore, that a disconnect can emerge between experts and their evidence as the latter is made relevant for policymaking. This disconnect is at the heart of understanding processes of politicisation and scientisation that occur during adaptation policymaking and is a key characteristic for understanding the tensions between expert and political authority for policymaking in liberal democratic government.

A conceptual framework for understanding climate change policy evidence

Based on these descriptions of the available evidence for adaptation, and building on the work of Jasanoff and Wynne (1998), I therefore propose a conceptual framework for understanding the epistemological character of the scientific evidence for climate change. Rather than a homogeneous expert evidence base that is relayed linearly to policymaking, as espoused by linear-technocratic policy models, adaptation policy evidence is derived, I argue, through a multi-component deliberative process (see Figure 4.1). Climate experts develop a body of trans-scientific evidence in the form of climate change models about the future trajectory of global and regional climates. This evidence is based on consensus expert understandings of the planet's bio-geophysical systems, the anthropogenic influences on them and a range of assertions and expert assumptions about how those influences will affect climate in the future (Petersen, 2012; Edwards, 2010). Importantly, this trans-scientific component is developed within the climate modelling community itself, with minimal input from a broader range of experts or non-expert evidence users.

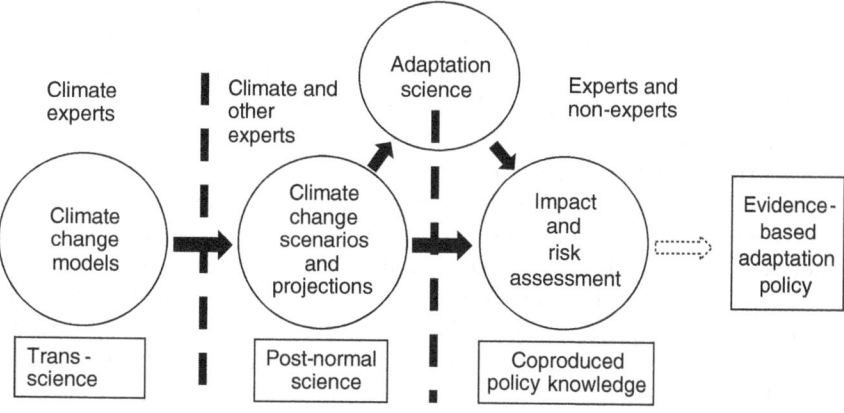

Figure 4.1 Knowledge co-production for adaptation policy (Tangney, 2016)

Second, as described by Hulme and Dessai (2008), a process of post-normal scientific development begins whereby climate experts negotiate with other expert and lay policy players the means by which climate change model outputs will be presented and portrayed to make them most meaningful for users of this information. Third, as Preston et al. (2013) describe, the research community coordinates the development of impact, receptor or vulnerability-specific information about potential harm. Whereas post-normal climate science involves the development of climate change scenarios and projections, adaptation science involves the development of other forms of science (e.g., impact models) to help understand the potential hazards and impacts from climate change.

Fourth, a process of knowledge coproduction ensues whereby those negotiated climate change model outputs and related adaptation science are used by expert and lay policy players to assess the likely impacts from climate change on government policy. This knowledge coproduction process may incorporate both the outputs of adaptation science and/or climate projections and scenarios directly. What results from this final stage is a body of internally consistent coproduced and negotiated policy knowledge to inform the political executive. In the cases described in Chapter 7, this knowledge coproduction process played out through a combination of technical modelling, 'expert elicitation' workshops, and a process of negotiation and review between experts, technical contractors and bureaucratic policy players about what constituted potential (or at least politically acceptable) impacts from future climate.

Under this model, and in the context of the linear-technocratic schema, a significant challenge to the legitimacy of evidence-based policy appears to be to define just what constitutes impartial (or at least credible and politically acceptable) evidence for climate adaptation policy. If the evidence used by policymakers is coproduced between expert and lay policy players through layers of normative decision-making, deliberation and consensus formation, this suggests that politics, overt or otherwise, has a role to play in the generation of policy evidence (Preston et al., 2013; Wesselink and Hoppe, 2011; Owens et al., 2004); a contribution that has largely gone unrecognised by the political executive (and indeed by many climate experts) in their pursuit of evidence-based policy or the rationalist linear-technocratic heuristics prescribed by government, and one that calls into question an ability to 'correctly' define and characterise adaptation impacts and risks.

A significant problem arises, I argue, because governments tend to assume that adaptation problems are discrete, bounded issues that can be objectively understood and that decision-makers can make effective pre-emptive decisions based on the linear provision and use of sound scientific evidence (see for example, DEFRA, 2011a, 2009; NEMC, 2010; APSC, 2007). In reality, however, adaptation problems often defy this rationalist logic (Adger et al., 2011; Head, 2008a; Reilly and Schimmelpfennig, 2000). Although a broad and uneasy consensus may be reached about the strategic character of many adaptation problems, I argue that they are, nonetheless, contingent on the perspectives provided by the level and scale of governance from which any given policy player or group of players seeks to address them.

The nature of climate adaptation policy problems

As discussed above, our understanding of the nature, scale and timing of future potential climate change hazards is, and likely always will be, partial and uncertain (Frigg et al., 2013a; Mearns, 2010). Scientific outputs describing these hazards provide no definitive answers and can be difficult for decision-makers to interpret and to use (Heaphy, 2015; Mullan et al., 2013; Dessai et al., 2009; Hall, 2007). The planet's bio-geophysical and social-ecological systems are highly complex, interdependent and only partially understood in many cases (Scheffer, 2009; Folke, 2006), so much so that we have a limited understanding of the potential scale, frequency and severity of future climate change impacts on those systems, and where thresholds and tipping points may exist (Schneider, 2004). The complex, non-linear dynamics of planetary systems means that it is also unwise to use past trends and present characteristics as the basis of predictions of future impacts (Adger et al., 2011). These uncertainties mean that climate change and adaptation science is of a variety that is often problematic when used to produce the types of policy evidence provided by *ex ante* appraisal such as cost-benefit, risk and impact assessments that seek empirically derived 'correct' answers to questions concerning what any particular problem is, how it should be resolved, and why (see for example, Willows and Connell, 2003). Nonetheless, climate change evidence is often conceptualised in terms of risk (see for example IPCC, 2014: p. 3).

The difficulties of climate change risk assessment

For the purposes of government policy appraisal, climate risk has, to date, often been assessed on the basis of two key components: evidence describing potential *consequences* of anthropogenic climate forcing, and estimates of the relative *likelihood* of those consequences at a given point in the future (UKCIP, 2009; Australian Government, 2006; Willows and Connell, 2003). Jones and Preston (2011) elucidate this risk assessment ideal further by explaining the difference between prescriptive and/or 'top-down' approaches to risk assessment that attempt to assess the likelihood and consequences of particular climate change hazards and their impacts on the basis of climate change projections and scenarios; versus diagnostic and/or 'bottom up' approaches that seek to understand existing vulnerabilities for individual jurisdictions or receptors and on that basis to assess risks in terms of achieving existing goals and objectives or exceeding thresholds.

When seeking to understand climate risks, however, the various uncertainties relating to climate, Earth system and social-ecological science highlighted above mean that discrete assessments of risk in terms of likelihood and consequence become difficult to produce. This has been described in the past in terms of navigating cascading uncertainties associated with our understanding of: GHG emissions → atmospheric GHG concentrations → global climate processes → regional climate responses → local climate responses → impacts on any given receptor (New et al., 2007; Moss and Schneider, 2000). This uncertainty cascade is shown in Figure 4.2, albeit in amended form to include the often overlooked uncertainties associated with the analysis and interpretation of climate change risks and the effectiveness of any derived policy responses (see Chapters 7 and 8).

Thus, we begin to see the problem with top-down or prescriptive risk-based approaches to understanding climate change that assume that risks can be

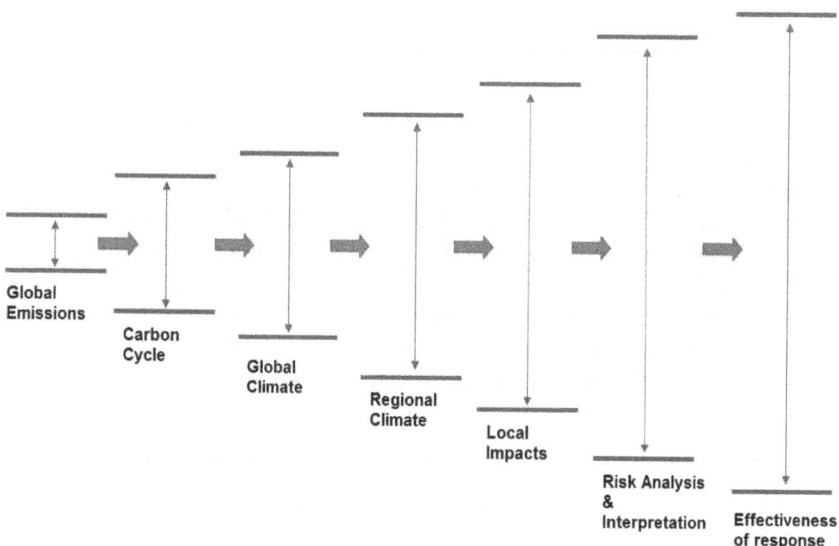

Figure 4.2 The uncertainty cascade, adapted from Moss and Schneider (2000: p. 7)

reliably derived. Empirical or semi-empirical risk assessment – the product of likelihood and consequence – as instructed by linear-technocratic policymaking rhetoric and guidelines (DEFRA, 2011a; NEMC, 2010; DEFRA 2009; Willows and Connell, 2003) is particularly problematic. Probabilities associated with the risks to a particular governance scale or level cannot be objectively assessed. This is due to the problem that frequentist estimations of the likelihood of occurrence of specific weather events or climatic states (i.e. probabilities determined on the basis of the frequency of past climate events) appear increasingly invalid due to the non-stationarity of an anthropogenically forced climate system, though they may still be worthwhile over the short-to-medium term. The assessment of probabilities are, therefore, heavily dependent on methods requiring subjective evaluation, either through degrees of congruence of a given outcome with climate model output (so-called Bayesian probabilities[5]), or on the basis of experts' and decision-makers' supposedly apolitical judgement (IPCC, 2007a; Dessai and Hulme, 2004; Grubler and Nakicenovic, 2001).

Neither of these subjective assessments of probability by the scientific community, however, have provided the types of answers sought by policymakers when using empirical risk assessment methods. The Bayesian techniques used for subjective probabilistic assessment of climate outcomes cannot provide discrete probabilities of future climatic states (UKCIP, 2009). These techniques are hindered by significant uncertainties relating to GCM downscaling techniques used to understand climate hazards at regional and local scales (Frigg et al., 2013a; Mearns, 2010; Dessai et al., 2009). In the UK, for instance, downscaled GCM output in the form of probabilistic projections are expected to be used to elicit policy preferences in relation to *ranges* of probable climatic states, yet have been hindered by considerable limitations in their use[6] (UKCIP, 2009; Hall, 2007, Dessai and Hulme, 2004). GCMs also cannot wholly account for the non-linearity of potential outcomes from anthropogenic forcing on the climate system, casting some doubt on the very scope of Bayesian probability distributions when accounting for potential outcomes (Frigg et al., 2013a; Schneider, 2004). Meanwhile, subjective judgements used in the semi-empirical assessment of likelihood of occurrence of potential future climate impacts are rife with difficulties, as highlighted in Chapters 6 and 7 for the case of UKCCRA. These relate to the contingency of expert perspectives, the power dynamics prevalent in 'expert elicitation' techniques used to reach consensus, and the prevalence of both intrinsic and extrinsic political influences in the assessment of risks.

The difficulties of understanding climate change impacts across governance scales

Irrespective of the knowledge limitations for understanding climate change hazards and their likelihood described above, the complexity and interconnectivity of social-ecological systems also makes it difficult to adequately understand the nature and scale of resultant impacts within a given jurisdiction, or to objectively define the boundaries of any adaptation problem. Global environmental problems with localised impacts such as climate change are relevant to, and require management across varying scales and levels of governance (Huitema et al., 2016; Cash et al., 2006; Cash and Moser, 2000). Effective policies (and presumably policy evaluation also) therefore require integration across levels since any particular policy may be apparently successful for one individual, organisation or government level, while simultaneously ineffectual or maladaptive for another (Corfee-Morlot et al., 2011; Urwin and Jordan, 2008; O'Brien et al., 2004). Conversely, risk analysis derived from single problem framings is likely to overlook significant characteristics of any particular adaptation problem. These difficulties also highlight the limitations of alternative vulnerability-based approaches to risk assessment. Cash et al. (2006) highlight the paradoxical nature of environmental assessment and management across levels of governance in this regard.

Top-down problem formulations and policy implementation, those often designed at national or regional levels, may be too blunt and insensitive to local conditions and constraints. Meanwhile, bottom–up approaches are insensitive to strategic considerations or larger social-ecological problems and are not easily generalisable (see Figure 4.3). One potential resolution to the limitations of risk-based policy appraisal is the utilisation of a two–way method that could combine top-down and bottom–up approaches (Jones and Preston, 2011), facilitated by governance networks that could bridge the interpretive gap between local and strategic governance levels and their contrasting under-standings of climate risks for adaptation policymaking (Howes et al., 2015).

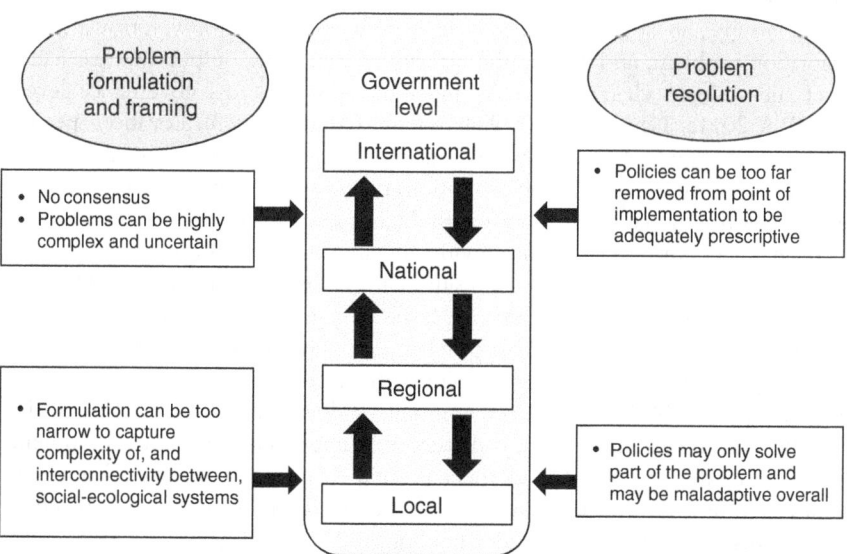

Figure 4.3 Formulating and resolving adaptation problems across scales and levels

However, such approaches are dependent on adequate integration and communication across governance levels; an attribute that has been sorely lacking in both Australian and UK policymaking jurisdictions for many years (see Chapter 3). As discussed in Chapter 8, this approach to risk assessment would also be as susceptible to a lack of clarity over the normative underpinnings of risk assessment as top-down or bottom-up approaches are on their own. Moreover, this two-way approach may still struggle with the technical limitations of incorporating climate science or producing risk assessments that align appropriately and usefully with any particular governance level, due to a mismatch of scale between climate science and climate vulnerability.

The difficulties of accounting for varying contextual understandings of potential climate impacts and risks are often compounded by a mismatch of scale between climate impacts and how climate change hazards are described through projections and scenarios. There is often a significant discrepancy in evidence provision between how we understand potential climate change hazards using the available scenarios and projections, and how we need to interpret and respond to climate risks in the process of evidence-based policymaking (Wilbanks and Kates, 1999). The resolution of GCMs/ESMs and their regionalised derivatives are often incompatible with the scale at which climate expresses itself and the location-specific characteristics of adaptation problems. In Chapter 6 I demonstrate that this is a principal reason for why climate change model outputs lack salience for many decision-makers.

These difficulties raise further significant questions about linear-technocratic policymaking approaches that rely upon risk assessment and cost-benefit analyses that overly depend on the available climate change science (or at least try to),[7] or that depend upon mono-scalar characterisations of climate risk. These methods largely assume an ability to objectively delimit and thus clearly formulate any adaptation problem, and assume the veracity of modelling outputs at a particular level and scale to characterise risks and direct policy across governance levels (DEFRA 2011a; NEMC, 2010; Willows and Connell, 2003). Location-specific risks, however, may be entirely different to those suggested by top-down risk assessments or by climate change projections that only provide hazard information at global and regional levels. Increasingly higher resolution projections may only increase uncertainty, even while giving the impression of doing the opposite (Dessai et al., 2009; Hall, 2007). And since climate change impacts are unlikely to fall neatly within existing governance levels (on geographical, jurisdictional or temporal scales) (Young, 2002), so problem definition, evaluation and resolution seem largely dependent on the viewpoint of the decision-maker rather than being objectively reducible by the expert community (Adger et al., 2011; Roome, 2001).

What this means, ultimately, is that existing approaches to the development of policy evidence often fail to meet the expectations of linear-technocratic policymaking rhetoric that advocates for the development of robust objective assessments of evidence for understanding and managing climate risks. The difficulties of reconciling contrasting yet equally valid problem framings across multiple governance levels, appear to make claims toward the apprehension of 'fact' or objective reality in the formulation and understanding of policy problems seem rather fanciful. Assessments of climate risks from a single governance level are necessarily subjective and contingent and, therefore, likely to be influenced by inevitable political inequalities present between policy players across levels and types of governance during stakeholder participation in policy evidence coproduction. And yet, as I describe for the cases of Queensland and the UK, governments have often constructed and promoted such assessments as fulfilling linear-technocratic expectations of policy evidence, and as being objective, transparent and true.

Climate adaptation's wicked characteristics

In order to understand the potential for political influence upon policy evidence for climate adaptation it seems sensible at this point to take a step back from the technical characteristics of climate science and policy evidence and to look at the problem through a theoretical framework relevant to many other public policy problems. The characteristics of both the science and the policy of adaptation may be usefully conceptualised through an understanding of what makes a policy problem 'wicked'. Rittel and Webber (1973) assigned wicked policy problems ten attributes which, they argue, make them remarkably difficult to adequately understand, address and to resolve:

1   Wicked problems have no agreed or definitive formulation.
2   Wicked problems have no stopping rule or end point (since the process of solving a wicked problem runs concurrent to the process of understanding its nature).
3   Solutions are not true-or-false, but good-or-bad (in other words, there are no agreed criteria to assess a 'correct' response).
4   There is no immediate or ultimate test of a solution to a wicked problem, since responses have potential for numerous unforeseen consequences.
5   Every response to a wicked problem is consequential and leaves impacts that cannot be undone.
6   There is no describable number of possible solutions to a wicked problem since there are no criteria to enable one to establish that all possible solutions have been identified.
7   Every wicked problem is essentially unique and therefore there is no suitable precedent to guide decision-makers.
8   Every wicked problem is a symptom of, or interconnected with, some other problem.
9   The choice of explanation of a wicked problem determines the nature of the problem's attempted resolution.
10   The decision-makers have no right to be wrong, since they are liable for the responses they generate, and mistakes in either action or inaction can be very costly.

As so succinctly described by Prins (2011), wicked problems such as climate change are 'open problems' – they are so complex and interconnected with everything else that we cannot know all the things we need to know in order to be able to fix them, nor can we know when we have enough knowledge about these problems to start fixing them. In this section I frame adaptation problems in terms of their wickedness,[8] to shed light on the difficulties faced by the policymaking community and to demonstrate how politics can contribute to the development of both adaptation evidence and policy.

Wicked problems have no agreed or definitive formulation

The differing characteristics of the adaptation policy problem, as exemplified by the two cases used for this research (see Chapter 3), attest to its susceptibility to divergent policy formulations. For policymakers, adaptation problems can be interpreted in a variety of ways that focus on resilience, vulnerability or exposure to climate, and they can be framed in terms of anthropogenic climate change, or not. For instance, they may be considered in terms of an ongoing necessity for managing society's exposure to climate variability in the present and the future without fully considering any increasing susceptibility to harm as portrayed by climate change projections, as in the case of Queensland. Alternatively, adaptation problems may be constructed as a new challenge requiring the development or redesign of institutional and governance structures to shore-up societal vulnerability to the looming threat of climate change, as in the UK. Which of these policymaking perspectives is more *objectively* correct is not possible to say, irrespective of the reality of the greenhouse effect and existing climate change. As I demonstrate in this book, these contrasting perspectives depend on the contextual political and socio-economic factors of a given jurisdiction and the normative/political priorities of government. Even from the perspective of experts and those in agreement about the threat of climate change, however, within a given jurisdiction or geographical area adaptation problems can be formulated differently and therefore may look considerably different depending upon one's perspective or area of expertise (Huitema, 2016; Sarewitz, 2004).

For instance, as described in Chapters 6 and 7, experts involved in the assessment of risks for the UK's climate change risk assessment had in some cases markedly different assessments of risk to those described in the final consensus of a national assessment. Likewise, in terms of both climate and adaptation science, varying assumptions and interpretations of uncertainties are required to understand the characteristics and dynamics of social-ecological systems. How environmental problems are constructed by experts is subject to a myriad of varying yet equally valid interpretations and uses of available sets of evidence. This is what Sarewitz (2000) refers to as the 'excess of objectivity' that occurs for evidence-based environmental policymaking. As discussed above in relation to the trans-scientific character of climate science, and for climate change projections specifically, the contingency of problem formulations may also be determined to some extent by the value judgements taken during the derivation of climate change science. This is demonstrated by Hulme et al. (2009) who used a number of equally valid climatic baselines in climate model runs to produce significantly different climate futures.

Box 4.1 The contingent nature of adaptation problems: The case of the disappearing village of Happisburgh in southeast England

On the southeast coast of England,* the village of Happisburgh sits along a coastline that, at its worst point, is eroding at a rate of 12 metres per year. A policy of 'managed retreat' or 'managed realignment' has been implemented by government for coastal communities such as Happisburgh, given that no amount of infrastructure is said to be realistically (or at least, cost-effectively) able to prevent the combined forces of coastal erosion and sea-level rise. Indeed, attempts at providing such infrastructure in the past have exacerbated erosive forces elsewhere along the coast (Adger et al., 2011). Therefore, these coastal communities face the potential for literal obliteration in the coming century since government will no longer maintain existing infrastructure under the Shoreline Management Plan for this location, nor provide any financial compensation to property owners from the destructive forces of the sea (North Norfolk District Council, 2012). Unsurprisingly, residents of

Figure 4.4 Happisburgh coastal erosion (CC BY-SA 2.0)

Happisburgh are not necessarily happy with the government's policy priorities and objectives, their interpretation of the associated risks from coastal erosion and climate change, nor their proposed policy response to them (BBC, 2014a, 2012a; Whiteside, 2014). This case is useful for demonstrating the contingent nature of adaptation problems such that their framing, problem formulation, risk characterisation and 'best' resolution depend on the perspective of any given stakeholder.

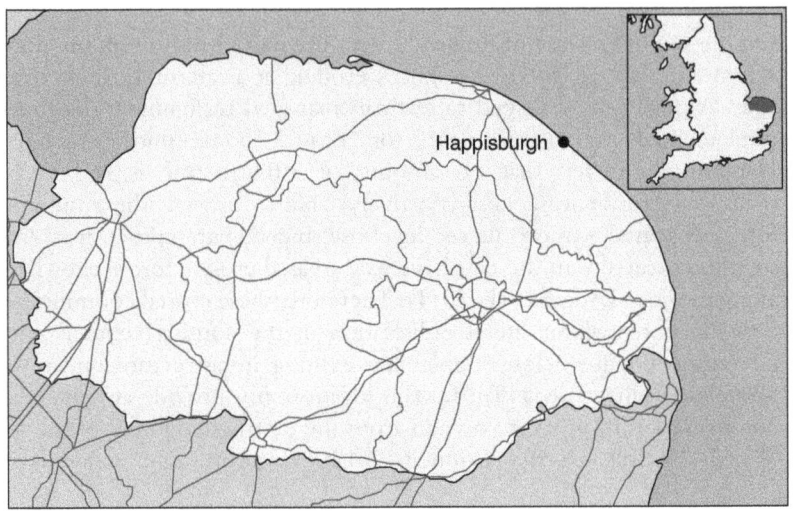

Figure 4.5 Happisburgh, Anglia, southeast England (CC BY-SA 3.0)

* Happisburgh is located in Anglia, a region of southeast England, though not located within the specific governance jurisdiction of Southeast England.

Wicked problems have no stopping rule or end point (since the process of solving a wicked problem runs concurrent to the process of understanding its nature)

Climate adaptation is a process of managing the risks presented by a complex, open and dynamic natural system and its interaction with the socio-economic systems of contemporary society. Government and society will always be subject to the vagaries of climate and thus will always have to adapt (Strauss and Orlove, 2003; Glacken, 1990). However, the prospect of anthropogenic climate change presents even greater uncertainty and risk, and because experts will never be able to provide definitive answers about the future, so the task of adapting must occur despite this uncertainty and alongside ongoing attempts to better understand the characteristics of our changing climate. As is well understood by the political science community, however (see Lindblom, 1959), it is in the nature of policymaking more generally that attempts to resolve policy issues often run concurrent to policy players' developing goals and priorities, as well as to their adequate identification and understanding of those issues.

Further, just as adaptation problems have no definitive formulation, they also have no definitive end point in terms of spatial or temporal scale. For example, according to current projections, sea level may continue to rise for many centuries to come, yet policy players in the UK, for instance, have chosen to delimit this problem for the southeast of England to the year 2100. Where and how policymakers delimit adaptation problems are often decisions based on normative/political priorities and/or the bounds of existing governance jurisdictions that are unlikely to align with the social-ecological and bio-geophysical bounds of any adaptation problem (Young, 2002) (see Boxes 4.1 and 4.3). In this regard, however, adaptation policy problems are similar to many other multi-level (though arguably far less wicked) policy issues.

Solutions are not true-or-false, but good-or-bad (in other words, there are no agreed criteria to assess a 'correct' response)

Adaptation is, in part, a process of prioritisation in the use of scarce public resources, using the best available evidence (Howes et al., 2015). Governments and communities must decide where to invest resources and to answer essentially political questions relating to the resilience 'of what and for whom'.[9] Furthermore, the uncertainties relating to climate change and our understanding of bio-geophysical and social-ecological systems as described above mean that we can never have unequivocal answers about what a correct response to climate risk may be. As Adger et al. (2011, p. 758) notes:

> Responses to one risk alone may inadvertently undermine the capacity to address other stressors, both in the present and future.

As such there are no correct adaptation actions, only good or bad ones depending on one's perspective on any given adaptation problem. Even the use of standard public policy criteria such as appropriateness, effectiveness and efficiency requires normative choices that cannot be assessed in terms of objective 'correctness'.

Box 4.2 Failing policy and evidence legitimacy: Seeking financial support for infrastructure improvements in SEQ

Although the Queensland Reconstruction Authority* (QRA) follows the mantra 'Build it back better' (QRA, 2011), referring to an intention to make damaged infrastructure more resilient to future extreme climate events, the influence of partisan politics has complicated the achievement of this aim in practice. Alongside conservative governments' dislike for climate *change* adaptation policy (see Chapter 3), there has also been an ongoing financial disincentive for the QRA to consider any future increase in the severity and frequency of extreme events associated with climate change. Under the National Disaster Relief and Recovery Arrangements (NDRRA) guidelines, federal government provides financial support to state government to rebuild infrastructure on a like-for-like basis, except for essential public assets where the state can demonstrate the cost-effectiveness of enhanced measures and their ability to mitigate future natural disasters (Australian Government, 2012). According to one interviewee involved in administering these state funding applications to the Commonwealth Government, what this means in effect is that most efforts to adapt to future climate change through the enhancement of infrastructure must be financed by state government alone (Aus-Policy Player 2).

This disincentive appears to reflect the lack of bipartisan legitimacy for climate science in Queensland, and the ongoing prioritisation for the development of a business case to inform policymaking, whereby climate science lacks sufficient political acceptability to influence the case for the cost-effectiveness of climate *change* adaptation measures. It is no surprise, therefore, that the QRA's Strategic Plan makes no mention of climate change (QRA, 2014). Moreover, the Newman government (2012–2015)

Figure 4.6 Queensland Reconstruction Authority's 2011 strategy (Queensland Government, 2011)

handed most responsibility for adaptation planning to local government[†] (Rickards et al., 2014; Preston et al., 2013), who have the fewest financial resources and least technical capacity to assess climate change risks. What this shifting of responsibilities and financial burden means, in effect, is that there is also little or no political space to incorporate climate science into policymaking or to consider climate change strategically across the region (Aus-Policy Players 2, 3).

[*] The QRA is a temporary government organisation set up in response to flooding events across the State of Queensland in 2010/2011.
[†] Through changes to the Local Government Act 2009 (Queensland Government, 2014a).

There is no immediate or ultimate test of a solution to a wicked problem, since responses have potential for numerous unforeseen consequences

Due to the complexity of climatic and bio-geophysical systems and the resulting intractable uncertainties surrounding future climate (Stocker et al., 2013), as well as the complex and uncertain dynamics of social-ecological systems (Scheffer, 2009), it is often not possible to adequately understand the extent to which any given adaptation response is likely to be effective in the future, or indeed whether it may ultimately prove beneficial or detrimental to communities in the long run. As Adger et al. (2011, p. 758) note, system complexities raise significant challenges in designing effective adaptation responses:

> Dealing with specific risks without full accounting of the nature of system resilience leads to responses that can potentially undermine long-term resilience.

Indeed, given these uncertainties, it may be argued that what constitutes a good adaptation response in the present relates to prevailing political and normative views and may inevitably be subject to *hindsight bias* by those who will have to live with the ultimate consequences. Priorities may have changed by the time it becomes possible to understand the efficacy of any adaptation response and, if past events are any indication (see Heazle et al., 2013), pre-emptive adaptation decisions may not be judged on the basis of the information that was available at the time they were made. In this regard, adaptation measures are similar to many other types of policy responses made under deep uncertainty. Yet, as discussed in Chapter 8, 'good' adaptation decisions are also strongly influenced by contrasting normative concepts of climate resilience and what we perceive as normality or the status quo.

Every response to a wicked problem is consequential and leaves impacts that cannot be undone

This characteristic is particularly relevant to pre-emptive climate change adaptation actions that may tie communities and their governments into particular types of response. For example, the construction of large-scale infrastructure may lead governments down a path of continually attempting to eliminate climate change risks through more and better infrastructure defences, or through the balancing of priorities in the operation of a given infrastructure asset, and thus, potentially neglecting other indicators of climate resilience such as adaptive capacity and social capital or 'social capacity' within communities, that may be required for effective climate hazard response (Kuhlicke et al., 2011). Levin et al. (2010) explain that care is needed when designing and implementing new policy approaches that may confine future adaptation activity into pathways that are difficult to diverge from. As exemplified by the Thames Estuary 2100 project in southeast England (EA, 2012b) (see Box 4.3), and the conceptual work of Heazle et al. (2013) in relation to disaster risk management in southeast Queensland, however, this type of policy *path-dependence* may be minimised or avoided through careful hedging strategies. These would attempt an evidence-based 'wait and see' approach involving careful planning and adaptive management to delay risky capital investment until absolutely necessary. Nonetheless, whether an adaptation response involves the building of infrastructure, or the restructuring of policymaking arrangements (e.g., the UK's introduction of the Climate Change Act (2008)), adaptation policy responses can leave lasting impacts upon policymaking and the communities served by it that can be difficult to undo. It is, perhaps in part, for this reason that governments in both the UK and Queensland have been reluctant to make specific infrastructure investments on the basis of uncertain climate change projections.

There is no describable number of possible solutions to a wicked problem since there are no criteria to enable one to establish that all possible solutions have been identified

As described in Chapter 8, given the uncertainties surrounding the trajectory of future climate, the task of adaptation is open to innumerable possible solutions which are likely to be more or less effective depending on the ultimate outcomes of climate change over the coming decades and stakeholders' conceptions of what it means to be climate resilient. Adaptation options are chosen, I argue, not just on the basis of the legitimacy given to alternative sets of available evidence, or the legitimacy for adaptation policymaking more generally, but also in relation to policymakers' risk management strategies which relate to underlying norms and priorities about how they value the present over the future, and how they seek to preserve what is valuable to them. Again, it is worth noting that this supposedly wicked characteristic is evident in many, if not most policy problems and, therefore, may be considered a lesser characteristic within Rittel and Webber's (1973) framework.

Every wicked problem is essentially unique and therefore there is no suitable precedent to guide decision-makers

Although national and state governments often attempt to address adaptation strategically through generic policy frameworks and evidence – for instance, using global or regional climate projections for national assessments of risk – it is an innate feature of environmental management issues such as climate adaptation that many of their characteristics are determined by local and contextual circumstances. Although governments often need to consider natural resource management and infrastructure planning at strategic governance levels, climate vulnerabilities are also unique to any given locale, community or jurisdiction, dependent on their geography, economy and the features of their particular social-ecological systems (Adger et al., 2011). Adaptation problems are also unique within any given locale because of how climate manifests itself at individual locations, which can vary widely in ways not accounted for by climate change projections or strategic climate impact assessments (Wilbanks and Kates, 1999; New et al., 2007). This means that although adaptation responses in other locations may provide useful sources of learning, most adaptation problems are essentially unique and cannot rely on a one-size-fits-all approach to policymaking.

Every wicked problem is a symptom of, or interconnected with, some other problem

Adaptation problems tend to be defined in terms of governance jurisdiction rather than according to their physical bounds (if such bounds could ever be clearly defined). They often relate to the behaviour of overlapping social-ecological and economic systems which means that the characteristics of an adaptation problem encompass a range of inter-related issues associated with the vulnerability and resilience of communities (Adger et al. 2011, p. 758).

In the UK, for instance, adaptation problems relate as much to the vulnerabilities created by the urban development and over-population of particular locations (such as southeast England) as they do to the severity of existing or future climate events (see Chapter 3). Similarly, in southeast Queensland adaptation problems are interconnected with problems of population growth, urban development, disaster risk and natural resource management as well as the vagaries of existing climate variability.

Box 4.3 Effective 'hedging' strategies for climate change adaptation: The case of the Thames Estuary in southeast England

The city of London and surrounding areas adjacent to the Thames Estuary in southeast England are dependent on a series of barriers, walls and embankments to protect this region from the combined threats of fluvial and coastal flooding, and storm surges (which the estuary is prone to as a result of low pressure systems arriving from the North Sea). Unlike flood risk and water resource management provisions for the city of Brisbane (described in Box 4.4 below), these flood defences are currently fit for purpose. The risks from climate change, however, mean that these defences are unlikely to continue to be adequate over the coming century. The Environment Agency of England & Wales has devised a plan (known as the Thames Estuary 2100 (TE2100) Project) that seeks to ensure that improvements and enhanced defences will be provided as and when they are required, rather than investing in a precautionary approach in advance of climate change risks that may not ultimately materialise. This adaptive management approach uses a 'hedging' strategy that considers the existing capacity of the estuary's flood defences and identifies key decision points throughout the century at which investment decisions will have to be made. At such points government will have to decide, based on the evidence available at the time, whether to maintain existing systems or to invest in new and/or replacement infrastructure (EA, 2012b). This decision-making process seeks to balance the optimality of adaptation measures with their robustness to uncertainty when managing flood risk. It essentially seeks to navigate an adaptation response between avoiding major pre-emptive infrastructure expenditure on the one hand, and on the other, being prepared for all but sudden climatic changes that may cause existing defences to become inadequate before there is time to replace or upgrade them.

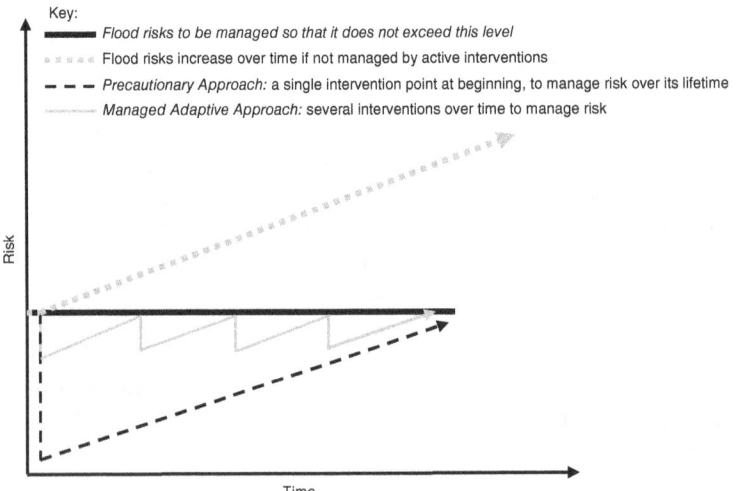

Figure 4.7 Managing flood risk through the century using the TE2100 managed adaptive approach (adapted from EA, 2012b)

The choice of explanation of a wicked problem determines the nature of the problem's attempted resolution

This characteristic is related to the issue of problem formulation described above. How an adaptation issue is explained depends on the perspectives of those experts and other policy players responsible for the construction of adaptation evidence. In turn this explanation and how policymakers frame it as a problem formulation, determines how they attempt to resolve it.

As described in Chapter 3, in Queensland the framing of climate adaptation aligns with existing policy objectives relating to disaster risk management and an ongoing need for community resilience to climate extremes, as well as to statist–developmentalist traditions of urban planning and natural resource management that prize the provision of infrastructure to overcome the state's ongoing exposure to climate extremes. Thus adaptation is often pursued both through the reduction or elimination of present risks and through the enhancement of the resilience of communities that are persistently exposed to climate extremes, with limited consideration of future climate change.

By contrast, in the UK, adaptation is pursued explicitly in terms of climate *change* adaptation, a risk that has, as yet, largely failed to materialise (or at least, the causal linkages between recent climate extremes and climate change remain largely inconclusive). In the UK climate change is perceived through abstract notions of future risk as described by climate projections and scenarios and through government's understanding of the UK's considerable vulnerability to climate. This framing of the adaptation problem has resulted in their principal focus being on the bolstering of institutional and governance arrangements to ensure the future resilience of socio-economic systems in the event of future extreme climate change (e.g., through the requirements of the Climate Change Act (2008)). These measures are, I argue, largely prompted by the intense vulnerabilities of the UK's over-populated and heavily urbanised society, which means that even relatively minor climate events can have a dramatic impact.

Infrastructure provision in the UK has, to date, been largely divorced from the kind of frequent and contrasting extreme events of the kind often experienced in Australia (Head et al., 2014). Although this may have begun to change in recent years as extreme events have become more frequent (BBC, 2014b), risk management infrastructure exists in the UK as much a result of its highly vulnerable urbanised society as due to its experiences of extreme events of the type seen in Queensland, or as anticipated as a result of climate change. And yet, although risk management may be prompted more by societal vulnerability than by persistent exposure to extremes, due to the legitimacy provided to climate change in the UK this infrastructure is being gradually upgraded to cope with future climate change (EA, 2012b). Nonetheless, very few, if any, infrastructure projects have been designed or built solely as a result of climate change, and the UK still lacks the experience and therefore, arguably, the expertise that Queensland has developed in terms of the practice of disaster risk reduction and emergency management. Queensland's expertise means that it appears to have made considerable progress in terms of enhancing a 'social capital' associated with community resilience to climate (QFCI, 2011).

The contrasting perspectives of exposure and vulnerability which frame these cases' approach to adaptation policy provide a useful conceptual distinction for understanding the contingent nature of adaptation problems and their resolution, and also why (as described in Chapter 7) the scientisation of adaptation policy proceeds by differing paths between these two cases.

Box 4.4 The dangers of rationalism and path-dependency: Operation of the Wivenhoe and Somerset dams in southeast Queensland

Both flood risk and water supply in the Brisbane area of SEQ are managed via the Wivenhoe and Somerset dams on the Brisbane River, which are required by policymakers to fulfil dual and often conflicting roles in the face of contrasting weather events. Flood risk management capacity is restricted to facilitate water storage for supply, or vice versa, which effectively reduces dam management to a zero-sum game between water security and flood mitigation. In the event of anticipated flooding, pre-emptive reductions in water levels below the dams' 'full supply level' – to allow space for storing flood waters and thus mitigate extreme flooding – increases vulnerability to water shortages during drought by reducing supply (QFCI, 2011). Decisions made about the appropriate use of the dams during extreme events necessarily involve, therefore, political decisions about which hazard presents a greater threat, particularly in the absence of the data needed to reliably forecast the risk of flooding versus drought (Heazle et al., 2013).

Such difficult decisions involving necessary trade-offs in the management of climate-related risk highlight the dangers of relying on rationalist 'predict-then-act' approaches to climate adaptation in the face of intractable uncertainties concerning future weather extremes and climate change. In actuality, such approaches require political prioritisation about competing risks and (potentially) vulnerable groups and locations that no amount of evidence or adherence to strict management protocols can avoid. Further, this case highlights the dangers of path-dependency that can result from a reliance on infrastructure to eliminate climate risk, whereby the construction of assets such as the Wivenhoe Dam can tie governments into policy prescriptions that can be very difficult to deviate from. In the case of SEQ, in the absence of further considerable capital investment, the city of Brisbane is now dependent on infrastructure that can provide neither optimal nor adequately robust risk management in the face of unpredictable and contrasting climate extremes (Tangney, 2015). Policy decisions in relation to flood risk and water resources management for Brisbane are dependent on the management of these dams and the zero-sum game played in balancing competing risks.

Figure 4.8 The Wivenhoe Dam, SEQ (CC – Public Domain)

The decision-maker has no right to be wrong, since they are liable for the responses they generate and mistakes in either action or inaction can be very costly

This characteristic relates to many contemporary policy problems given the impracticalities and political barriers of attempting a trial and error approach to policymaking. For adaptation this difficulty is compounded by adaptation policies' attempts to manage interconnected socio–economic and ecological systems which mean that failed adaptation policies may have many unforeseeable effects. However, politically, the importance of policymaking accountability may be even more prevalent in the UK case as a result of the Climate Change Act (2008) which ensures that climate change policies must be considered beyond the timescales of the political electoral cycle. There is, therefore, added impetus on policymakers to ensure that they don't tie themselves into ineffective policy actions that cannot be undone at a later stage.

Levin et al. (2010) identified four further characteristics with which they characterised climate change as a 'super wicked problem'. Although this characterisation was primarily aimed at the broader issue of climate change, these criteria also directly relate to the specific task of climate adaptation.

Time is running out

Given the intractable uncertainties associated with understanding future climate (Stocker et al., 2013), it is possible that changes may occur swiftly and with little warning as a result of unforeseen tipping points in the Earth's bio-geophysical systems. This possibility is important in light of the timescales required to construct certain types of disaster risk management or climate adaptation infrastructure. The ability of governments to provide appropriate risk management may be constrained by such time limits. An example from the UK in this regard relates to plans for the construction of flood defences for the city of London and surrounding areas of SEE (see Box 4.3). The Thames Estuary 2100 project was prompted by a realisation by the Environment Agency of England & Wales, London City Council and their partners that any upgrade to existing flood defences would require time to construct, time which may not be available in the event of a rapid climatic change causing an increase in sea-level rise, storm-surge events, fluvial flooding or some combination of all three hazards. Such climatic changes would likely overcome existing defences, since they are already at the limit of their capacity. This project sought to develop a plan whereby flood defences could be progressively improved in line with the expected impacts indicated by climate change models, thus ensuring that the increased pressure on flood defences on the Thames from climate change could be managed in a timely manner, without unnecessarily or prematurely tying government into significant investment should it not be required (EA, 2012b).

There is no central authority with responsibility to resolve the problem

In both Australian and UK cases, the difficulties encountered when managing adaptation problems across jurisdictional borders relate to the difficulties described above and in Chapter 3 of governing environmental problems across governance scales and levels. In Australia, this can be seen in relation to the interactions within and between states when managing cross-border environmental issues such as water management for the Murray Darling river basin (Steele et al., 2012). Moreover, federal government has passed most adaptation responsibilities to state government, who in turn appear to have left most responsibility for adaptation policy to local government councils (Preston et al., 2013). In circumstances where adaptation problems transcend state borders, there often appears to be no central authority with responsibility to provide authoritative governance, depending instead on the achievement of some form of political consensus to be reached by participating state policy players.[10] In the face of anthropogenic climate change, this could potentially result in considerable difficulties for the achievement of coherent adaptation responses.

Similarly, in the UK, much like for the issue of acid rain in the 1970s (see Chapter 3), adaptation issues that transcend sovereign borders can cause difficulties for neighbouring member states of the EU, where there is little political or economic incentive for one party to act. Although the possibility for such issues to arise in relation to the UK's climate adaptation may be limited by its geography, since the UK is an island state, it is nonetheless possible to foresee significant trans-boundary issues relating to, for example, climate change-induced refugees and immigration.

Those seeking to solve the problem are also causing it

In relation to adaptation problems, this characteristic speaks to the theoretical work of Beck (1992) and Giddens (1990) relating to the reflexive modernisation of contemporary society, whereby the trappings of modernity have increased the vulnerability of society to climate-related events. Socio-economic priorities of government relating to economic growth, increasing urbanisation and rapidly growing populations have, in many cases, exacerbated problems such as flooding, drought, deteriorating water quality, biodiversity loss and bushfires. These difficulties are due to both the form and location of associated urbanisation and densification that have increased the vulnerability of communities to climate events. As such, it may be argued that governments seeking to remedy problems associated with climate variability and change are also largely responsible for the policy and planning decisions that exacerbate such problems. This is particularly applicable to the case-studies examined in this book since both southeast England and southeast Queensland are rapidly growing regions with an increasingly urbanised landscape.

Decision-makers and the public tend to make decisions that reflect very short time horizons and therefore discount the future beyond what is required to solve a long term problem like climate change

This is an important issue in relation to climate adaptation as with other areas of policymaking, but also concerning the legitimacy of policy evidence relating to the potential future impacts of climate change. As will be described in further detail in Chapter 6, one of the principal reasons why climate science lacks legitimacy and salience for adaptation decisions is that it does not provide information over timescales relevant to political and policy decision-making. The levels of uncertainty associated with the available evidence mean that, although often perceived as credible, in practice this evidence lacks legitimacy and/or salience relative to more deterministic information concerning a business case for short term policy action. Even when considering investment decisions over the longer term, commonly accepted discount rates used for cost-benefit assessment and aligning with economic-rationalist principles will preference short term priorities over longer ones. This short-termism, in combination with an accompanying lack of salience/legitimacy of climate science, is compounded by a mismatch between decision-making timescales over which adaptation problems must be considered and the timescales of the political electoral cycle.

The aforementioned 'wicked' and 'super wicked' characteristics of climate adaptation problems mean that both climate science and subsequent policy evidence are heavily constrained when it comes to their ability to meet linear-technocratic expectations for the objectivity of expert knowledge or its ability to linearly inform decision-making. As we shall see in later chapters, adaptation policy problems are extremely difficult to address and resolve in a way that is acceptable to all, or even most, policy players and stakeholders. The development, presentation and use of policy evidence requires subjective, normative choices concerning the framing and characterisation of climate risks and the use of evidence for policymaking.

Conclusion

Climate adaptation presents an intriguing problem for evidence-based policymaking. It is at once both a highly contingent problem that demands layers of value judgement by both experts and non-experts in order to understand it, while also being dependent on supposedly objective expert authority for its technical credibility and political legitimacy as an extant policy priority. Adaptation problems must be considered by governments at both local and strategic scales, and the tendency for varying formulations and framings of these problems depending on the level of government at which it is considered, suggests that norms, values and even political deliberation have a role to play, not just in policy development, but in the evidence used to inform and legitimise that policy.

In this chapter, I have shown how climate science is just one component in a complex process of policy evidence development for climate adaptation. As this science comes closer to government decision-making, processes of interpretation and coproduction are ultimately necessary since, in order to be salient and legitimate as well as credible, evidence must account for the norms and values of decision-makers and the applications for which they seek to use it. As I discuss further in Chapter 7, failure to balance these evidence attributes results in the non-use of that evidence.

This chapter demonstrates how the entire suite of problems associated with climate adaptation seems to present a series of difficulties that makes them exceedingly difficult to impartially define, understand and to resolve. Adaptation's wicked characteristics and its contingent nature strongly suggest the potential for political influence in how adaptation problems are understood during evidence development. These wicked characteristics also raise important questions about the extent to which adaptation evidence can maintain legitimacy across levels of government or across the bipartisan political divide while maintaining credibility and salience to ensure it is both useful and usable for policymaking. In the next chapter, I discuss some of the theory behind expert-political interactions that can help to understand how wicked characteristics manifest themselves in policy practice.

Notes

1   Though even here there can be some disagreement. See for example, the arguments made by Curry (2013).
2   'interpreting of evidence in ways that are partial to existing beliefs, expectations, or a hypothesis in hand' (Nickerson, 1998: p. 175).
3   The exception being the provision of GHG emissions scenarios that aid climate experts' understanding of future climate change. Known as Representative Concentration Pathway (RCP) scenarios in the IPCC's 5 Assessment Report, they are derived using Integrated Assessment Models including economic, demographic, energy and simple climate components (Stocker et al., 2013, pp. 79, 80). RCPs constitute a significant trans-scientific component of climate change modelling.
4   Swart et al. (2014) make an important distinction between science *for* adaptation (relating to understanding the impacts from and vulnerabilities to climate change) and science *of* adaptation (which investigates and challenges how we think about adaptation). Although a useful distinction, its discussion lies somewhat outside the remit of this book. When referring to adaptation science, I am principally referring to science *for* adaptation, while I hope this book contributes to the science *of* adaptation.
5   See Chapter 6, note 6 for a more detailed description. Bayesian probabilities have a subjective component to their derivation, based on a prior assumed probability of occurrence, which is then updated as new observations are made (or, in the case of climate change, as model runs are conducted) (Bertsch McGrayne, 2011).
6   For instance, in their first iteration the UK Climate Projections 2009 allowed such probabilistic sets to be used for individual 25 km^2 grid squares but warned against aggregating or averaging of adjacent grid squares in order to understand potential impacts for a user-defined area (UKCIP, 2009).
7   In practice, as described in Chapters 6 and 7, this has proven to be a rather difficult task.
8   One of the most thoughtful commentators on the concept of wicked policy problems, Prof. Brian Head at the University of Queensland, has argued that wickedness has become something of a catch-all phrase for policy problems and as a result has lost much of its potency as an operational concept in distinguishing between those that are particularly intractable. I'm inclined to agree; most contemporary policy problems display at least some wicked characteristics. With this in mind, I use Rittel and Webber's (1973) criteria here as a means of distinguishing what is unusual about adaptation problems, but also what is similar to many other problems faced by policymakers in order to highlight the challenges in developing evidence-based responses to climate adaptation.
9   See Chapter 8 for a discussion of the concept of climate resilience and its application to policymaking.
10   Although tentative steps toward such coordination have been made through the Council of Australian Government's Climate Change Adaptation Framework (COAG, 2007).

Bibliography

Adger, WN, Brown, K, Nelson, DR, Berkes, F, Eakin, H, Folke, C, Galvin, K, Gunderson, L, Goulden, M, O' Brien, K, Ruitenbeek, J and Tompkins, EL 2011, 'Resilience implications of policy responses to climate change', *WIRES Climate Change*, vol. 2, no. 5, pp. 757–766.
Australian Government 2006, *Climate Change Impacts and Risk Management: A Guide for Business and Government*, Australian Greenhouse Office, Department of Environment and Heritage. Canberra, ACT, Australia.

Australian Government 2012, *Natural Disaster Relief and Recovery Arrangements: Determination Version 1*, Commonwealth of Australia, Attorney General's Department, Canberra, viewed 27 May 2014, www.disasterassist.gov.au/NDRRADetermination/Pages/default.aspx/.

Australian Public Service Commission [APSC] 2007, *Tackling Wicked Problems: A Public Policy Perspective*, APSC, Australian Government, Canberra.

BBC 2012a, 'Living with coastal erosion in Happisburgh, East Anglia', British Broadcasting Corporation, viewed 20 September 2012: www.bbc.co.uk/learningzone/clips/living-with-coastal-erosion-in-happisburgh-east-anglia-pt-1-2/7361.html/.

BBC 2014a, 'Britain from Above: Happisburgh', British Broadcasting Corporation, viewed 24 August 2014, www.bbc.co.uk/learningzone/clips/living-with-coastal-erosion-in-happisburgh-east-anglia-pt-1-2/7361.html/.

BBC 2014b, 'Met Office: Evidence "suggests climate change link to storms"', British Broadcasting Corporation, 9 February, viewed 10 February 2015, www.bbc.com/news/uk-politics-26084625/.

Beck, U 1992, *Risk Society: Towards a New Modernity*, Sage, London.

Bertsch McGrayne, S 2011, *The Theory that Would Not Die: How Bayes' Rule Cracked the Enigma Code, Hunted Down Russian Submarines, & Emerged Triumphant from Two Centuries of Controversy*, Yale University Press, London.

Cash, DW and Moser, SC 2000, 'Linking global and local scales: Designing dynamic assessment and management processes', *Global Environmental Change*, vol. 10, no. 2, pp. 109–120.

Cash, DW, Clark, WC, Alcock, F, Dickson, NM, Eckley, N, Guston, DH, Jager, J and Mitchell, RB 2003, 'Knowledge systems for sustainable development', *Proceedings of the National Academy of Sciences*, vol. 100, no. 14, pp. 8086–8091.

Cash, DW, Adger, WN, Berkes, F, Garden, P, Lebel, L, Olsson, P, Pritchard, L and Young, O 2006, 'Scale and cross-scale dynamics: Governance and information in a multi-level world', *Ecology and Society*, vol. 11, no. 2, viewed 16 March 2015, www.ecologyandsociety.org/vol10/iss2/art9/.

Clark, WC and Majone, G 1985, 'The critical appraisal of scientific inquiries with policy implications', *Science, Technology & Human Values*, vol. 10, no. 3, pp. 6–19.

Corfee-Morlot, J, Cochran, I, Hallegatte, S and Teasdale, P 2011, 'Multilevel risk governance and urban adaptation policy', *Climatic Change*, vol. 104, no. 1, pp. 169–197.

Council of Australian Governments [COAG] 2007, *National Climate Change Adaptation Framework, Council of Australian Governments*, Commonwealth of Australia, viewed 9 September 2012, www.climatechange.gov.au/government/initiatives/national-climate-change-adaptation-framework.aspx/.

Curry, J 2013, 'Consensus distorts the climate picture', *The Australian*, 21 September 2013.

Department of Environment, Food and Rural Affairs [DEFRA] 2009, *Adapting to Climate Change: Helping Key Sectors to Adapt to Climate Change*, DEFRA, London, viewed 11 March 2013, http://archive.defra.gov.uk/environment/climate/documents/interim2/report-guidance.pdf/.

Department of Environment, Food and Rural Affairs [DEFRA] 2011a, *Greenleaves 3: Guidelines for Environmental Risk Assessment and Management*, DEFRA, London, viewed 17 March 2015, www.defra.gov.uk/publications/files/pb13670-green-leaves-iii-1111071.pdf.

Department of Environment, Food and Rural Affairs [DEFRA] 2012a, *UK Climate Change Risk Assessment: Government Report*, HMSO, Norwich, UK, viewed 6 May 2013, www.defra.gov.uk/publications/files/pb13698-climate-risk-assessment.pdf.

Department of Environment, Food and Rural Affairs [DEFRA] 2012b, *UK Climate Change Risk Assessment: Evidence Report*, DEFRA, viewed 6 May 2013, http://randd.defra.gov. uk/Default.aspx?Menu=Menu&Module=More&Location=None&Completed=0&Pro jectID=15747#RelatedDocuments/.

Dessai, S and Hulme, M 2004, 'Does climate adaptation policy need probabilities?', *Climate Policy*, vol. 4, no. 2, pp. 107–128.

Dessai, S, Hulme, M, Lempert, R and Pielke Jr, R 2009, 'Climate Prediction: A Limit to Adaptation?', in Adger, WN, Lorenzoni, I and O' Brien, KL (eds), 2009, *Adapting to Climate Change: Thresholds, Values, Governance*, Cambridge University Press, Cambridge.

Douglas, H 2007, 'Rejecting the Ideal of Value-Free Science', in Kincaid, H, Dupre, J and Wylie, A (eds), *Value-Free Science? Ideals and Illusions*, Oxford University Press, Oxford.

Douglas, H 2009, *Science, Policy, and the Value-Free Ideal*, University of Pittsburgh Press, Pittsburgh, PA.

Edwards, PN 2010, *A Vast Machine: Computer Models, Climate Data and the Politics of Global Warming*, The MIT Press, London.

Environment Agency of England & Wales [EA] 2012b, *Thames Estuary 2100. Managing Flood Risk through London and the Thames Estuary: TE2100 Plan*, EA, viewed 30 July 2014, www.gov.uk/government/uploads/system/uploads/attachment_data/file/ 322061/LIT7540_43858f.pdf.

Fagan, B 2000, *Floods, Famines and Emperors*, Pimlico, London.

Fischer, F 2009, 'Technical Knowledge in Public Deliberation: Towards a Constructivist Theory of Contributory Expertise', in Fischer, F, *Democracy and Expertise: Reorienting Policy Inquiry*, Oxford University Press, Oxford.

Folke, C 2006, 'Resilience: The emergence of a perspective for social-ecological systems analyses', *Global Environmental Change*, vol. 16, no. 3, pp. 253–267.

Frigg, R, Bradley, S, Du, H and Smith, LA 2013a, *Laplace's Demon and Climate Change, Centre for Climate Change Economics and Policy Working Paper No. 121*, Grantham Research Institute on Climate and the Environment, viewed 14 May 2013, http://seamusbradley. nfshost.com/Papers/demon-grantham.pdf.

Frigg, R, Smith, LA and Stainforth, DA 2013b, 'The myopia of imperfect climate models: The case of UKCP09', *Philosophy of Science*, vol. 80, no. 5, pp. 886–897.

Frigg R, Smith LA and Stainforth DA 2015. 'An assessment of the foundational assumptions in high-resolution climate projections: The case of the UKCP09', *Synthese*, vol. 192, no. 12, pp. 3979–4008.

Funtowicz, SO and Ravetz, JR 1993, 'Science for the post-normal age', *Futures*, vol. 25, no. 7, pp. 739–755.

Giddens, A 1990, *The Consequences of Modernity*, Polity Press, Cambridge.

Glacken, TJ 1990, *Traces on the Rhodian Shore: Nature and Culture in Western Thought from Ancient Times to the End of the Eighteenth Century*, 5th edn, University of California Press, Berkeley, CA.

Grubler, A and Nakicenovic, N 2001, 'Identifying dangers in an uncertain climate', letter submitted to *Nature*, vol. 412, no. 6842, pp. 15.

Hall, J 2007, 'Probabilistic climate scenarios may misrepresent uncertainty and lead to bad adaptation decisions', *Hydrological Processes*, vol. 21, no. 8, pp. 1127–1129.

Head, B 2008a, 'Wicked problems in public policy', *Public Policy*, vol. 3, no. 2, pp. 101–118.

Head, L, Adams, M, McGregor, HV and Toole, S 2014, 'Climate change and Australia'. *WIRES Climate Change*, vol. 5, no. 2, pp. 175–197.

Heaphy, LJ 2015, 'The role of climate models in adaptation decision-making: The case of the UK climate projections 2009', *Euro Jnl Phi Sci*, vol. 5, no. 5, pp. 233–257.

Heazle, M, Tangney, P, Burton, P, Howes, M, Grant-Smith, D, Reis, K and Bosomworth, K 2013, 'Mainstreaming climate change adaptation: An incremental approach to disaster risk management in Australia', *Environmental Science & Policy*, vol. 33, pp. 162–170.

Hertin, J, Turnpenny, J, Jordan, A, Nilsson, M, Russel, D and Nykvist, B 2009, 'Rationalising the policy mess? Ex ante policy assessment and the utilisation of knowledge in the policy process', *Environment and Planning A*, vol. 41, no. 5, pp. 1185–1200.

Hoppe, R 2005, 'Rethinking the science-policy nexus: from knowledge utilization and science technology studies to types of boundary arrangements', *Poiesis Prax*, vol. 3, no. 3, pp. 199–215.

Howes, M, Tangney, P, Reis, K, Grant-Smith, D, Heazle, M, Bosomworth, K and Burton, P 2015, 'Towards networked governance: Improving interagency communication and collaboration for disaster risk management and climate change adaptation in Australia', *Journal of Environmental Planning and Management*, vol. 58, no. 5, pp. 757–776.

Huitema, D, Adger, WN, Berkhout, F, Massey, E, Mazmanian, D, Munaretto, S, Plummer, R and Termeer, CCJAM 2016, 'The governance of adaptation: choices, reasons, and effects. Introduction to the special feature', *Ecology and Society*, vol. 21, no. 3, pp. 37.

Hulme, M and Dessai, S 2008, 'Negotiating future climates for public policy: A critical assessment of the development of climate scenarios for the UK', *Environmental Science & Policy*, vol. 11, pp. 54–70.

Hulme, M 2009, *Why We Disagree about Climate Change: Understanding Controversy, Inaction and Opportunity*, Cambridge University Press, Cambridge.

Hulme, M, Dessai, S, Lorenzoni, I and Nelson, DR 2009, 'Unstable climates: Exploring the statistical and social constructions of "normal" climate', *Geoforum*, vol. 40, no. 2, pp. 197–206.

Intergovernmental Panel on Climate Change [IPCC] 2007a, *Climate Change 2007: The Physical Science Basis. Contribution of Working Group I to the Fourth Assessment Report of the Intergovernmental Panel on Climate Change*, Cambridge University Press, Cambridge and New York.

Intergovernmental Panel on Climate Change [IPCC] 2014, 'Summary for policymakers', in Field, CB, Barros, VR, Dokken, DJ, Mach, KJ, Mastrandea, MD, Bilir, TE, Chatterjee, M, Ebi, KL, Estrada, YO, Genova, RC, Girma, B, Kissel, ES, Levy, AN, MacCracken, S, Mastrandea, PR and White, LL (eds), 2014, *Climate Change 2014: Impacts, Adaptation and Vulnerability. Part A: Global and Sectoral Aspects. Contribution of Working Group II to the Fifth Assessment Report of the Intergovernmental Panel on Climate Change.* Cambridge University Press, Cambridge and New York.

Ioannidis, JPA 2005, 'Why most published research findings are false', *PLoS Med*, vol. 2, no. 8, pp. 124.

Jasanoff, S and Wynne, B 1998, 'Science and decision-making', in Rayner, S and Malone, EL (eds), *Human Choice and Climate Change*, Volume 1: *The Societal Framework*, Batelle Press, Columbus, OH.

Jones, RN and Preston, BL 2011, 'Adaptation and risk management', *WIRES Climate Change*, vol. 2, pp. 296–308.

Jordan, A 2002b, 'Efficient Hardware and Light Green Software: Environmental Policy Integration in the UK', in Lenschow, A (ed.), 2002, *Environmental Policy Integration: Greening Sectoral Policies in Europe*, Earthscan, London.

Kellow, A 2009, *Science and Public Policy: The Virtuous Corruption of Virtual Environmental Science*, Edward Elgar Publishing, Cheltenham, UK.

Kuhlicke, C, Steinfuhrer, A, Begg, C, Bianchizza, C, Brundl, M, Buchecker, M, De Marchi, B, Di Masso Tarditti, M, Hoppner, C, Komac, B, Lemkow, L, Luther, J, McCarthy, S, Pellizoni, L, Renn, O, Scolobig, A, Supramaniam, M, Tapsell, S, Wachinger, G, Walker, G, Whittle, R, Zorn, M and Faulkner, H 2011, 'Perspectives on social capacity building for natural hazards: Outlining an emerging field of research and practice in Europe', *Environmental Science & Policy*, vol. 14, no. 7, pp. 804–814.

Levin, K, Cashore, B, Bernstein, S and Auld, G 2010, 'Playing it forward: Path dependency, progressive incrementalism, and the "super wicked" problem of global climate change', *International Studies Association Convention*, 28 February – 3 March 2010, Chicago, IL, viewed 17 March 2015, http://citation.allacademic.com/meta/p_mla_apa_research_citation/1/7/9/7/0/pages179707/p179707-1.php/.

Lindblom, CE 1959, 'The science of "muddling through"', *Public Administration Review*, vol. 19, no. 2, pp. 79–88.

Mearns, LO 2010, 'The drama of uncertainty', *Climatic Change*, vol. 100, no. 1, pp. 77–85.

Meyer, R 2011, 'The public values failures of climate science in the US', *Minerva*, vol. 49, no. 1, pp. 47–70.

Miller, CA and Edwards, PN 2001, *Changing the Atmosphere: Expert Knowledge and Environmental Governance*, The MIT Press, London.

Moss, RH and Schneider, SH 2000, 'Uncertainties in the IPCC TAR: Recommendations to Lead Authors for More Consistent Assessment and Reporting', in Pachauri, R, Taniguchi, T and Tanaka, K (eds), 2000, *Guidance Papers on the Cross-Cutting Issues of the Third Assessment Report of the IPCC*, World Meteorological Organisation, Geneva, pp. 39, viewed 16 January 2016, http://lib.riskreductionafrica.org/bitstream/handle/123456789/1143/177.The%20Third%20Assessment%20Report.%20Cross%20Cutting%20Issues%20Guidance%20Papers.pdf?sequence=1&isAllowed=y/.

Moss, RH, Meehl, GA, Lemos, MC, Smith, JB, Arnold, JR, Arnott, JC, Behar, D, Brasseur, GP, Broomell, SB, Busalacchi, AJ, Dessai, S, Ebi, KL, Edmonds, JA, Furlow, J, Goddard, L, Hartmann, HC, Hurrell, JW, Katzenberger, JW, Liverman, DM, Mote, PW, Moser, SC, Kumar, A, Pulwarty, RS, Seyller, EA, Turner II, BL, Washington, WM and Wilbanks, TJ 2013, 'Hell and high water: Practice-relevant adaptation science', *Science*, vol. 342, no. 6159, pp. 696–698.

Mullan, M, Kingsmill, N, Kramer, AM and Agrawala, S 2013, *National Adaptation Planning: Lessons from OECD Countries*, OECD Environment Working Papers No. 54, OECD Publishing, viewed 14 May 2013, http://dx.doi.org/10.1787/5k483jpfpsq1-en/.

National Emergency Management Committee [NEMC] 2010, *National Emergency Risk Assessment Guidelines*, Attorney General's Department, Commonwealth of Australia, Canberra.

New, M, Lopez, A, Dessai, S and Wilby, R 2007, 'Challenges in using probabilistic climate change information for impact assessments: An example from the water sector', *Phil. Trans. R. Soc. A.*, vol. 365, pp. 2117–2131.

Nickerson, RS 1998, 'Confirmation bias: A ubiquitous phenomenon in many guises', *Review of General Psychology*, vol. 2, no. 2, pp. 175–220.

Nilsson, M, Jordan, A, Turnpenny, J, Hertin, J, Nykvist, B and Russel, D 2008, 'The use and non-use of policy appraisal tools in public policymaking: An analysis of three European countries and the European Union', *Policy Science*, vol. 41, no. 4, pp. 335–355.

North Norfolk District Council 2012, *Shoreline Management Plan 6: Kelling Hard to Lowestoft Ness*, NNDC, viewed 24 August 2014, www.northnorfolk.org/smp6/SMP6_documents.html/.

O' Brien, K, Sygna, L and Haugen, JE 2004, 'Vulnerable or resilient? A multi-scale assessment of climate impacts and vulnerability in Norway', *Climatic Change*, vol. 64, nos. 1–2, pp. 193–225.

Oreskes, N 2004, 'Science and public policy: What's proof got to do with it?', *Environmental Science and Policy*, vol. 7, pp. 369–383.

Oreskes, N, Shrader-Frechette, K and Belitz, K 1994, 'Verification, validation and confirmation of numerical models in the earth sciences', *Science*, vol. 263, pp. 641–646.

Owens, S, Rayner, T and Bina, O 2004, 'New agendas for appraisal: Reflections on theory, practice, and research', *Environment and Planning A*, vol. 36, no. 11, pp. 1943–1959.

Petersen, AC 2012, *Simulating Nature: A Philosophical Study of Computer Simulation Uncertainties and Their Role in Climate Science and Policy Advice*, 2nd edn, CRC Press, Taylor & Francis Group, New York.

Pielke Jr, RA 2015, 'I am under "investigation"', *The Climate Fix: Various Musings on Climate Science and Policy*, 25 February 2015, viewed 19 December 2016, https://theclimatefix.wordpress.com/2015/02/25/i-am-under-investigation/.

Porter, J and Dessai, S 2016, 'Is co-producing science for adaptation decision-making a risk worth taking?', Centre for Economics and Climate Change Policy – Working Paper No. 263, Sustainability Research Institute – Paper No. 96, March 2016, viewed 19 December 2016, www.cccep.ac.uk/publication/is-co-producing-science-for-adaptation-decision-making-a-risk-worth-taking/.

Preston, BL, Rickards L, Dessai S and Meyer, R 2013, 'Water, Seas, and Wine: Science for Successful Climate Adaptation', in Moser, SC and Boykoff, MT (eds), *Successful Adaptation to Climate Change: Linking Science and Policy in a Rapidly Changing World*, Routledge, Abingdon.

Prins, G 2011, *The Wicked Problem of Climate Change*, viewed 10 September 2012, www.youtube.com/watch?v=Tiqrv5wMxuA/.

Queensland Floods Commission of Inquiry [QFCI] 2011, *Queensland Floods Commission of Inquiry: Interim Report*, QFCI, Brisbane, viewed 3 April 2013, www.floodcommission.qld.gov.au/publications/interimreport/.

Queensland Government 2014a, *Local Government Act 2009*, Queensland Government website viewed 17 December 2014, https://www.legislation.qld.gov.au/LEGISLTN/CURRENT/L/LocalGovA09.pdf.

Queensland Reconstruction Authority [QRA] 2014, *Strategic Plan 2014–15*, Queensland Government, Brisbane, viewed 11 February 2015, http://qldreconstruction.org.au/publications-guides/reconstruction-plans/.

Queensland Reconstruction Authority [QRA] 2011, *Rebuilding a Stronger, More Resilient Queensland*, Queensland Government, viewed 19 December 2014, http://qldreconstruction.org.au/u/lib/cms2/rebuilding-resilient-qld-full.pdf.

Radaelli, CM and Meuwese, ACM 2010, 'Hard questions, hard solutions: Proceduralisation through impact assessment in the EU', *West European Politics*, vol. 33, no. 1, pp. 136–153.

Reilly, J and Schimmelpfennig, D 2000, 'Irreversibility, uncertainty, and learning: Portraits of adaptation to long-term climate change', *Climatic Change*, vol. 45, no. 1, pp. 253–278.

Rickards, L, Wiseman, J and Edwards, T 2014, 'The problem of fit: scenario planning and climate change adaptation in the public sector', *Environment and Planning C: Government and Policy*, vol. 32, no. 4, pp. 641–662.

Rittel, HW and Webber, MM 1973, 'Dilemmas in a general theory of planning', *Policy Sciences*, vol. 4, no. 5, pp. 155–169.

Roome, N 2001, 'Conceptualising and studying the contribution of networks in environmental management and sustainable development', *Business Strategy and the Environment*, vol. 10, no. 2, pp. 69–76.

Sarewitz, D 2000, 'Science and Environmental Policy: An Excess of Objectivity', in Frodeman, R (ed.), *Earth Matters: The Earth Sciences, Philosophy, and the Claims of Community*, Prentice Hall, Ann Arbor, MI.

Sarewitz, D 2004, 'How science makes environmental controversies worse', *Environmental Science & Policy*, vol. 7, no 5, pp. 385–403.

Scheffer, M 2009, *Critical Transitions in Nature and Society*, Princeton University Press, Princeton, NJ.

Schneider, SH 2004, 'Abrupt nonlinear climate change, irreversibility and surprise', *Global Environmental Change*, vol. 14, no. 3, pp. 245–258.

Steele, W, Sporne, I, Shearer, S, Singh-Peterson, L, Serrao-Neumann, S, Crick, F, Dale, P, Low Choy, D, Eslami-Endargoli, L, Iotti, A and Tangney, P 2012, *Learning from Cross-Border Arrangements to Support Climate Change Adaptation in Australia: Stage 1*, Urban Research Program, Research Monograph 14, Griffith University, Brisbane, Australia.

Stocker, TF, Qin, D, Plattner, GK, Alexander, LV, Allen, SK, Bindoff, NL, Bréon, FM, Church, JA, Cubasch, U, Emori, S, Forster, P, Friedlingstein, P, Gillett, N, Gregory, JM, Hartmann, DL, Jansen, E, Kirtman, B, Knutti, R, Krishna Kumar, K, Lemke, P, Marotzke, J, Masson-Delmotte, V, Meehl, GA, Mokhov, II, Piao, S, Ramaswamy, V, Randall, D, Rhein, M, Rojas, M, Sabine, C, Shindell, D, Talley, LD, Vaughan, DG and Xie, S-P 2013, 'Technical Summary', in *Climate Change 2013: The Physical Science Basis. Contribution of Working Group I to the Fifth Assessment Report of the Intergovernmental Panel on Climate Change*, Cambridge University Press, Cambridge.

Strauss, S and Orlove, B (eds) 2003, *Weather, Climate, Culture*, Berg, Oxford.

Swart, R, Biesbrook, R and Capela Lourenco, T 2014, 'Science of adaptation to climate change and science for adaptation', *Frontiers in Environmental Science*, vol. 2, no. 29, pp. 1–7.

Tangney, P 2015, 'Brisbane City Council's Q100 Assessment: How climate risk management becomes scientised', *International Journal of Disaster Risk Reduction*, vol. 14, no. 4, pp. 496–503.

Tangney, P 2016, 'The UK's 2012 climate change risk assessment: How the rational assessment of science develops policy-based evidence', *Science and Public Policy*, published online 6 September 2016, https://academic.oup.com/spp/article-abstract/doi/10.1093/scipol/scw055/2525558/The-UK-s-2012-Climate-Change-Risk-Assessment-How/.

Turnpenny, J, Radaelli, CM, Jordan, A and Jabob, K 2009, 'The policy and politics of policy appraisal: Emerging trends and new directions', *Journal of European Public Policy*, vol. 16, no. 4, pp. 640–653.

UK Climate Impacts Program [UKCIP] 2009, *UKCP09: UK Climate Projections*, Environmental Change Institute, University of Oxford, UKCIP website, viewed 9 September 2012, http://ukclimateprojections.defra.gov.uk/22676/.

Urwin, K and Jordan, A 2008, 'Does public policy support or undermine climate change adaptation? Exploring policy interplay across different scales of governance', *Global Environmental Change*, vol. 18, no. 1, pp. 180–191.

Waldman, S 2017, 'Judith Curry Retires, citing "craziness" of climate science', *E&E News*, 4 January 2017, viewed online 9 January 2017, www.eenews.net/stories/1060047798/.

Walker, KJ 2002, 'Environmental Policy in Australia', in Desai, U (ed.), 2002, *Environmental Politics and Policy in Industrialized Countries*, MIT Press, Cambridge, MA.

Weinberg, AM 1972, 'Science and trans-science', *Minerva*, vol. 10, no. 2, pp. 209–222.

Wesselink, A and Hoppe, R 2011, 'If post-normal science is the solution, what is the problem?: The politics of activist environmental science', *Science, Technology & Human Values*, vol. 36, no. 3, pp. 389–412.

Whiteside, J 2014, *Happisburgh Village Website: Coastal Erosion at Happisburgh*, viewed 24 August 2014, www.happisburgh.org/ccag/.

Wilbanks, TJ and Kates, RW 1999, 'Global change in local places: How scale matters', *Climatic Change*, vol. 43, no. 3, pp. 601–628.

Willows, R and Connell, R (eds) 2003, *Climate Adaptation: Risk, uncertainty and decision-making*, UK Climate Impacts Programme, Oxford.

Young, OR 2002, *The Institutional Dimensions of Environmental Change: Fit, Interplay, and Scale*, The MIT Press, Cambridge, MA.

5 Knowledge systems for sustainability

> Expertise is not merely something that is in the heads and hands of skilled persons, constituted through their deep familiarity with the problem in question, but rather … it is something acquired, and deployed, within particular historical, political, and cultural contexts.
>
> Sheila Jasanoff

Knowledge Systems is a term sometimes used to describe the institutional arrangements and interactions that seek to harness science, technology and other forms of expert knowledge for policy decision-making. In two influential[1] papers, Cash et al. (2002, 2003) suggested that effective knowledge systems for sustainable development can be created by managing the boundaries between knowledge and action through effective communication, translation and mediation. To achieve this boundary management, they propose a framework for assessing the effectiveness of evidence for use in decision-making using the criteria of *credibility*, *salience* and *legitimacy*. In this chapter I examine these concepts in some detail, in order to better understand the locus and character of the science-policy interface and how expertise and political authority interact at this boundary. I then examine Cash et al.'s formulation for what these interactions look like and to what extent their ideas can be helpful for climate adaptation scholars and practitioners. Since publication, their framework has been widely cited and continues to be used by those interested in the use of science for environmental policy.

Although in its original form it is limited in terms of its descriptive and prescriptive ability, I argue that the simplicity of Cash et al.'s framework makes it worthy of particular attention as a climate adaptation practitioner's model for understanding effective knowledge systems for policymaking. In particular, I believe it can, with some adjustment be a helpful conceptual device for scholars and policy players seeking to understand the tensions between expertise and politics during policymaking. This framework can also help to explicate the difficulties of adhering to linear-technocratic decision-making methods in practice. In this chapter, therefore, I propose an augmented form of Cash et al.'s framework which I then use in Chapter 6 to understand just how effective climate science and policy evidence has been in Queensland and the UK to date.

Effective evidence and experts in knowledge systems

As shown in Chapter 4 the facts concerning climate adaptation, including those about future climate change, are not determined by wholly objective means. Further, the development and content of policy evidence is not solely determined by the available scientific facts; considerations of *which* facts are most relevant and politically acceptable, as well as the norms and political machinations of the evidence development process, also play a part. Expert judgement, it would seem, is a necessary commodity. During policy evidence development, however, it seems that the problems of legitimacy and extension of expertise, as described by Collin and Evans (2002) (see Chapter 2), are addressed through an essentialist tactic of assigning 'expert' status to almost any policy evidence that is co-produced between experts (certified or otherwise) and non-experts. Although the democratisation of expertise – through knowledge co-production – for climate adaptation seems entirely appropriate, if not also necessary given its contingent nature, this evidence process appears particularly at risk from a lack of normative transparency when distinguishing the things we know, from information that simply aligns with our values and ideals.

The apparent mischaracterisation of the science-policy interface under essentialist, linear-technocratic principals, although politically expedient, does not satisfactorily resolve the problems of legitimacy and extension of expertise. This is because this demarcation tactic fails to adequately distinguish valid expert knowledge and judgement from the politics of policymaking. It does, however, highlight the potential problems of politicisation during evidence-based policymaking. Evidence development is influenced by how experts and other policy players interact and interpret the facts; the governance structures they are subject to; the policy and decision-making norms to which they must adhere; and, ultimately, by the prevailing political values and priorities of democratic government. While it is important not to conflate valid evidence with valid expertise (Jasanoff, 2003c), therefore, it is also important to account for their intimate relationship when policy evidence is necessarily infused with contingency and normative influence, while nonetheless portrayed as definitive and apolitical. The tendency to assign the same expert status to co-produced policy knowledge undertaken during policymaking, as to trans-scientific endeavours conducted within the academy, highlights the importance of balancing appropriate criteria of effective evidence to ensure it is sufficiently credible, useful and usable for policymaking.

To understand and manage the science-policy interface effectively, Cash et al. (2002, pp. 4–5) approach the problem of valid experts and useful evidence from the perspective of the knowledge production process and its perceived efficacy. They propose three subjective criteria by which to judge the effectiveness of knowledge processes for decision-making:

Credibility: 'whether an actor perceives information as meeting standards of scientific plausibility and technical adequacy';

Salience: the perceived relevance of information for an actor's decisions, or for the decisions that affect that actor; and,

Legitimacy: 'whether an actor perceives the process of knowledge production by the system as unbiased and meeting the standards of political and procedural fairness'.

These characteristics, they explain, are interdependent and must be balanced against one another. Efforts at enhancing one may diminish the other two. When addressed appropriately at the science-policy interface, effective knowledge systems can be designed in ways that minimise conflict between these criteria as much as possible while maintaining an adequate level of each. Traditionally, however, and as suggested by the work of Collins and Evans (2002) and their critics (described in Chapter 2), the focus on both sides of the science-policy interface has been on technical credibility as the principal determinant of useful, usable scientific knowledge and expertise for policy, often at the expense of its perceived salience and legitimacy.

This traditional focus on credibility can be seen, for example, in the collation of evidence by the Intergovernmental Panel on Climate Change (IPCC). Cash et al. (2002), citing Agrawala (1998), suggest that in the early stages of the development of IPCC outputs, stakeholders complained of a lack of inclusiveness of scientists and climate change modelling from developing countries. It appeared that considerations of evidence legitimacy were being neglected in favour of scientific credibility. Although the eventual inclusion of climate change models from developing countries expanded the range of uncertainty associated with the IPCC's conclusions and therefore may have diminished their credibility, the perceived legitimacy of the knowledge production process was ultimately enhanced for a wider range of users of IPCC outputs, thus enhancing its usefulness and political acceptability.

Likewise, although efforts were made to enhance the salience of the UK's Climate Projections 2009 (UKCP09) (Porter and Dessai, 2016), as I discuss in Chapter 6, this attribute may have been of secondary importance to the perceived credibility of that evidence as being the foremost source of climate change science publicly available in the UK (Tangney and Howes, 2016). Ultimately, however, UKCP09 was not perceived by many users to be particularly usable for decision-making (Tang and Dessai, 2012). Meanwhile, others have argued that Cash et al.'s (2003) criteria are insufficient on their own and further criteria, such as the iterativity of knowledge processes (when used as a normative criterion) and the political acceptability of the evidence produced (as a descriptive criterion), are also useful either to ensure effective evidence development or to account for political influences (Sarkki et al., 2015; Tangney and Howes, 2016).

Cash et al. (2003) conclude that society still does not know how to design effective knowledge systems due to the difficulties of balancing the competing needs for credibility, salience and legitimacy. Their analysis draws attention to a key difficulty encountered in the design of expert knowledge systems that is illustrated by the tensions and complementarities between these criteria, and that relates to the difficulties of finding an appropriate balance between expert and political authority (Irwin, 2006; Jasanoff, 2005; Collins and Evans, 2002). Namely, that in order to produce good policy evidence under a linear-technocratic schema, it is necessary to both preserve the privileged status of its expert authority by excluding political or normative influences as much as possible, while simultaneously engaging a range of non-expert (or non-certified expert) perspectives in its design and development, to ensure that evidence is legitimate and salient for those with political authority. This tension between expert and political authority during evidence development results in difficult trade-offs.

Cash et al.'s framework is important because it elucidates how the science-policy boundary, in practice, must be bridged by either scientist or non-expert policy player, or by an intermediary (i.e. a boundary player or organisation), in order to ensure the salience of the evidence produced within a legitimate knowledge system. In this way we can produce socially and/or policy-relevant science that is considered fair and representative of appropriate values. Too much emphasis on providing salient evidence in a legitimate way, however, may compromise the credibility of evidence since our perceptions of credibility are so often dependent on the impartiality of expertise, and upon assumptions about the importance of apolitical expert knowledge. Salient and legitimate climate change evidence, however, requires a variety of both expert and non-expert policy players to have a say in the underlying norms and values from which this evidence is derived. Meanwhile, too much emphasis on legitimacy through engagement with a wide range of stakeholders can compromise salience by producing evidence that fails to address the needs of any particular set of decision-makers. And, too much emphasis on salience may compromise credibility by failing to account for the nuance, complexity and uncertainties associated with the evidence describing the policy problems in question.

Given these conflicting needs for useful, usable knowledge, how then may we design effective evidence for adaptation policy? This seems a particularly daunting task given that the bounds of the science-policy interface – or where expert authority ends and political authority begins – in the development of climate adaptation policy evidence is so often difficult to specify, other than through the essentialist tactic of assigning all policy knowledge the label of 'expert evidence'. As suggested by critics of Collins and Evans' (2002) ideas about 'contributory expertise', in the context of providing effective knowledge about contingent but nonetheless technical policy problems, it is not easy to understand either who should be considered an expert or what balance of characteristics supposedly expert policy evidence should possess. And, as I demonstrate further in Chapters 6 and 7, only by balancing the technical and normative requirements of effective evidence in a transparent way can we also avoid processes of covert politicisation that can mask significant values and priorities within the provision of supposedly objective evidence, or avoid the subsequent suppression of normative debate through the scientisation of policymaking (Sarewitz, 2004).

Cash et al.'s (2002) framework highlights how policy players are inevitably forced to address an ongoing conflict of interest in the development, interpretation and use of evidence for policy. Both expert and non-expert policy players may be motivated to work on either side of the science-policy divide and appear to pursue, implicitly or otherwise, two opposing goals. Both experts and non-experts:

1   seek to separate expert and political authority in order to 'shore up their claims on cognitive authority' (Cash et al., 2002: p. 7); and simultaneously,
2   seek to bridge the science-policy divide in order to make useful, usable science, and/or to expedite normative priorities within an increasingly contested political space.

Bridging the science-policy gap is presumably achieved through negotiation between expert and non-expert policy players concerning user needs and the subsequent content of evidence (see Chapters 7 and 8). However, when we struggle to substantiate claims toward expert impartiality given the trans-scientific and post-normal character of climate and adaptation science, or even to identify who qualifies as a legitimate expert given the contingency of adaptation problems, this bridging endeavour adds weight to ideas concerning the politicisation of *ex ante* appraisal of evidence for policymaking, and compromises the validity of governments' claims to linear-technocratic policymaking (see Chapter 7).

The criteria of salience and legitimacy speak to a need for ensuring that both expert and non-expert policy players have a say in the underlying values from which evidence is derived (Maasen and Weingart, 2005) and speaks to Fischer's (2009) idea about the need for a socio-cultural as well as technical reasoning in technical decision-making in the public sphere (see Chapter 2). Yet, the credibility of policy evidence for climate change is still dependent on expectations concerning privileged expert knowledge and the immutability and objectivity of the associated science. For instance, climate change advocates in Australia actively promote the credibility of the associated certified expert consensus on global warming due to the (at times highly) contested political landscape surrounding this issue (see for example, Skeptical Science, 2015).

It would seem that effective knowledge systems should nonetheless seek to account for the norms and values that constitute a fundamental component of policy evidence, including answering questions about what constitutes legitimate expert judgement versus implicit normative prioritisation. In this regard, Collins and Evans' (2002) competing problems of the legitimacy and extension of expertise relate to our understanding of the interactions between the credibility, salience and legitimacy of effective knowledge systems. In its original form, however, Cash et al.'s framework only partially accounts for this inter-relation between political and expert authority. Although, in order to be effective, the knowledge production process must be perceived as sufficiently fair, unbiased and inclusive of appropriate values, concerns and perspectives, so too must the evidence produced. In other words, legitimacy is not just a characteristic of the knowledge production process or the experts involved, as suggested by Cash et al., it is also an important characteristic of evidence itself.

As I seek to show here, any discussion therefore concerning effective knowledge systems must account for a broader political acceptability of the available evidence, which is as important an attribute for effective knowledge systems as perceptions about its fair and unbiased production process. Cash et al.'s (2002: p. 1) paper explores 'how effective boundary work involves creating salient, credible, and legitimate information simultaneously for multiple audiences'. In its current form, however, their characterisation appears to involve a range of ambiguities and uncertainties relating to the interdependent relationships between their proposed criteria, as well as the interactions between knowledge production systems and the broader sphere of policymaking. Whilst interdependencies between these criteria are inevitable, some of the ambiguities that are largely left unaddressed by Cash et al. (2002, 2003) suggest a need to rethink their precise definitions. In particular, I argue, these criteria do not adequately address whether effective knowledge production can occur in isolation from the prevailing political forces that act upon evidence-based policymaking. In the following section, therefore, I discuss these three criteria in more detail and explain why, in order to fully account for the tensions between expert and political authority in evidence development and use, they require some adjustment to account for prevailing political priorities and ideals.

The importance of political acceptability for the legitimacy of effective knowledge systems

Let's begin by considering the concept of evidence salience. Cash et al.'s definition, I argue, does not fully elucidate the relationship between expert knowledge and political priorities for policymaking when considering this attribute. In a liberal democracy we expect appropriate expertise, generally speaking, to be relevant (i.e., salient) for use in a corresponding sphere of policymaking. The absence of salient expert advice could result in purely ideological or emotive decision-making that would suffer from a deficit of rationality and therefore of public acceptability (Weingart, 1999; Jasanoff, 1990). Nonetheless, granting experts a seat at the policymaking table does not mean that individual components of their evidence will be necessarily salient. Expertise and evidence will only be considered relevant if the knowledge they provide is politically acceptable. Salience for specific sets of evidence, therefore, will depend upon the political acceptability of particular policy issues with which that evidence aligns, or upon the acceptability of normative/political positions which are contained within, or ascribed to, that evidence by policy players. As Weingart (1999: p. 156) notes: 'If scientific knowledge is linked in any way to 'interests' (in policy-making), it is evaluated as supportive, contradictory, or even dangerous.'

In the case of climate adaptation policymaking, we can say that problem formulation and framing (e.g., in relation to the political acceptability for anthropogenic climate change – see Chapter 3) was a necessary precondition for the salience of climate change science in Australia and the UK. In turn, however, making evidence more salient through processes of 'post-normal' science, consultation and co-production not only enhances its instrumental utility, it may also contribute to enhancing the political acceptability of available knowledge and therefore enhance the legitimacy of associated policy issues. As McNie (2007: p. 20) notes: 'Production of salient, and thus relevant, information increases the likelihood that future decisions will be embraced.' Without knowledge of the framing of a policy issue like climate adaptation, and the political acceptability of experts' conclusions about this issue, I argue, it is impossible to fully understand or perceive the salience of evidence for a problem like climate adaptation.

Salience, as defined by Cash et al. (2002), however, is ambiguous because it does not adequately distinguish between the relevance of information in relation to government agenda-setting, versus its relevance in terms of the practical usability of that knowledge. Although Cash et al. (2002) did not fully elucidate any distinction (they hint at both concepts), to date salience has been interpreted and used to describe the practical, *instrumental* utility of evidence for informing the decision-making process, once a policy issue has already been identified and its framing is already agreed upon.[2] Under this interpretation, salience is dependent primarily on processes of consultation and negotiation between evidence producers and users to understand their information requirements (see for example, Porter and Dessai, 2016; Hulme and Dessai, 2008). This narrower concept of salience is entirely appropriate but, I argue, it is nonetheless dependent upon the prior and ongoing political acceptability of a given policy issue and, therefore, upon how it is framed by the political executive. This is because policy issues can often be framed in ways that make important sets of evidence irrelevant to begin with.

Weingart (1999: p. 157) envisages the issue of agenda-setting for policy issues with scientific underpinnings in terms of a 'recursive model' rather than the usual linear alternative of 'problem perception – expert advice – political decision'. This recursive model involves the following stages:

> Perception of the problem may come either from the scientific community or from policy-makers. In the political process it is transformed according to political criteria of relevance. As a political programme funding research for further clarification of the initial problem, it is handed back to the scientific community. The scientific community, in turn, executes the pertinent research whose results become the basis for continuous adaptation of the initial problem perception.
>
> (Weingart, 1999: p. 157)

Salience interpreted as a characteristic that encompasses only the latter (instrumental) part of this recursive model, therefore, seems somewhat limited in the context of understanding its relationship with the criteria of legitimacy and credibility, since it does not fully account for its relationship with a broader political acceptability for the underlying policy issue and its framing. Unfortunately, Cash et al. (2003) do not make an adequate distinction between these two types of information relevance.

Similarly, Cash et al.'s (2002) definition of legitimacy refers to the importance of *accounting for values, interests and specific circumstances from multiple perspectives within the knowledge production process*, and therefore appears to doff its cap to political influence and the need for political acceptability for the resultant evidence. However, this definition is worded in a way that confines this political influence to the production process itself and not also to perceptions of the knowledge produced. As a result, I argue, the concept of legitimacy as Cash et al. (2002) define it appears to overlap excessively with their concurrent definition of credibility, particularly in the context of post-normal co-design and bureaucratic co-production processes. If a knowledge production process is not fair or is thought to be biased, then presumably the resulting knowledge lacks technical credibility in the eyes of those who seek to use it. Both experts and technical evidence garner their technical credibility, at least in part, from their claims toward objectivity. Cash et al.'s (2002, 2003) distinction between credibility and legitimacy also appears, rather perversely, to lack usability itself, as demonstrated by its application in the studies of Weichselgartner and Kasperson (2010) and Tang and Dessai (2012), whereby, these terms have been commonly conflated and confused by interview participants asked about the utility of evidence for policy.

Cash et al. (2002, 2003) fail to explain why their description of legitimacy (as a characteristic of knowledge *production* rather than the knowledge itself) is important and different from the concurrent idea of credibility as a determinant of effective knowledge systems and evidence for policy, while the importance of political legitimacy (or acceptability) of the resulting evidence is neglected. One reason may be that it is neither easy, nor desirable, to purposefully design effective knowledge systems in a way that produces evidence that can always be both politically acceptable and wholly credible. As Dilling and Lemos (2011) argue, it is not possible for experts to control the context in which their evidence will be used. When using their framework as a set of normative (rather than descriptive) criteria, the idea that we should design evidence outputs to align with policymaking values also speaks to the politicisation of expert evidence. And yet, as I have shown in Chapter 4, all climate adaptation evidence contains a measure of value judgement in its derivation, so the strict rationalising function long presumed to be the role for evidence and expertise in policymaking does not apply with the same stringency as was once assumed. Inevitably there must be a weighing of values between those prevailing from deliberative democracy and those ascribed in, or attributed to, expert knowledge.

Cash et al.'s neglect of the political acceptability of evidence, however, may also be because they intended this criterion to be subsumed within the criterion of salience. Yet, as discussed above, this then leads to confusion between the political and instrumental salience of evidence. Further, Cash et al.'s focus may arise from a desire to understand the knowledge system and not simply the resulting knowledge. In respect to political legitimacy, however, these two components of evidence-based policymaking are interdependent and contingent upon the political context in which they exist, so descriptive or prescriptive criteria of effective knowledge systems must account for all three.

I argue that a broader concept of evidence legitimacy, incorporating a political acceptability for the evidence itself, can more effectively inform us about the tensions inherent between expert and political authority within liberal democratic government. This revised concept encapsulates and transcends issues concerning the fairness or unbiased nature of the knowledge production process. When conceived of as a broader political acceptability, the legitimacy granted to evidence for policymaking is derived from a valid combination of discernible expert judgement and authority inscribed during the knowledge process, and its congruence with the prevailing values and priorities of political authority (derived from democratic representation). If we assume that good policy decision-making (and therefore effective knowledge systems) is dependent upon some pragmatic combination of expert and political authority (Ezrahi, 1990), the absence of one form of authority will likely have adverse effects upon the acceptability of the other and therefore upon the acceptability of any subsequent decisions.[3]

I propose, therefore, the following revised definition of evidence legitimacy to help describe effective knowledge systems for climate adaptation policymaking:

Legitimacy: The degree to which information for policymaking is considered politically acceptable and the process of knowledge production considered to be unbiased and inclusive of appropriate values, interests, concerns, and specific circumstances from multiple perspectives.

This concept of legitimacy, as both a characteristic of evidence itself *and* the knowledge production process, is one that encapsulates a broader political acceptability for evidence in decision-making. This definition relates in important ways to evidence credibility since a lack of political acceptability for certain evidence sets may reflect in policy players' perceptions of evidence credibility, though not in a way that muddles their conceptual underpinnings; it merely speaks to a cognitive dissonance between facts and values (Kahan et al., 2012). This revised definition, I argue, is less ambiguous with concurrent understandings of credibility than Cash et al.'s (2002) original definition of legitimacy whereby they were essentially referring to the same attribute.

This revised definition of legitimacy also relates in important ways to the concept of salience; political acceptability is an important precursor for the political salience of particular tranches of expert knowledge (e.g., climate change science) as well as for subsequent endeavours to make that expert knowledge more instrumentally salient for policymaking. Thus, this revised definition also aligns usefully with Weingart's (1999) concept of a recursive model of agenda-setting for evidence-based policy. Importantly, however, this revised definition does not resolve the problems of politicisation prevalent in the development of expert evidence. Indeed if misused as a normative prescription, it could even exacerbate these difficulties. Nonetheless, as a descriptive model this revised form can help adaptation practitioners to elucidate these problems more effectively for the benefit of understanding the tensions and trade-offs between expertise and politics.

Conclusion

Discussion surrounding issues of demarcation between science and policy are complex. They can often descend into esoteric debates about the limits of expert authority in technical decision-making and the interplay between expertise and politics during policymaking. In this chapter I have sought to shed some light on the inevitable interactions that must occur between expert and non-expert knowledge and the socio-cultural value judgements required during the development of evidence for adaptation policymaking.

The task of balancing the needs for credible, legitimate and salient evidence in a way that maintains the influence of legitimate expert knowledge and judgement has proven to be highly challenging to date. Effective knowledge systems inevitably weigh the values inscribed during the development of policy evidence (and its scientific antecedents) with the values and priorities of prevailing politics. Past assumptions, about the ability of science and expertise to provide immutable truths in a way that can fit seamlessly into the considerations of policy decision-making, have never been accurate. As I describe in Chapter 6, however, it seems that these assumptions are also no longer satisfactory in the eyes of many policy players that encounter and seek to resolve complex and contentious policy problems, for which scientific understandings provide only partial answers.

Notes

1   At least, within the environmental sciences community. Cash et al.'s ideas concerning knowledge systems may have been considered somewhat simplistic by scholars of political science and Science & Technology Studies.
2   Note, for example, the interpretations made by Tang and Dessai (2012) and Wilbanks and Kates (1999) amongst others when highlighting the problems of developing relevant climate change science.
3   Since a lack of political authority would result in technocracy, and a lack of expert authority would result in purely ideological decision-making.

Bibliography

Cash, D, Clark, W, Alcock, F, Dickson, N, Eckley, N and Jager, J 2002, *Salience, Credibility, Legitimacy and Boundaries: Linking Research, Assessment and Decision Making*, John F. Kennedy School of Government, Harvard University, Faculty Research Working Papers Series, November 2002.

Cash, DW, Clark, WC, Alcock, F, Dickson, NM, Eckley, N, Guston, DH, Jager, J and Mitchell, RB 2003, 'Knowledge systems for sustainable development', *Proceedings of the National Academy of Sciences*, vol. 100, no. 14, pp. 8086–8091.

Collins, HM and Evans, R 2002, 'The third wave of science studies: Studies of expertise and experience', *Social Studies of Science*, vol. 32, no. 2, pp. 235–296.

Dilling, L and Lemos, MC, 2011, 'Creating usable science: Opportunities and constraints for climate knowledge use and their implications for science policy', *Global Environmental Change*, vol. 21, no. 2, pp. 680–689.

Ezrahi, Y 1990, *The Descent of Icarus: Science and the Transformation of Contemporary Democracy*, Harvard University Press, Cambridge, MA.

Fischer, F 2009, 'Technical Knowledge in Public Deliberation: Towards a Constructivist Theory of Contributory Expertise', in Fischer, F 2009, *Democracy and Expertise: Reorienting Policy Inquiry*, Oxford University Press, Oxford.

Hulme, M and Dessai, S 2008, 'Negotiating future climates for public policy: A critical assessment of the development of climate scenarios for the UK', *Environmental Science & Policy*, vol. 11, no. 1, pp. 54–70.

Irwin, A 2006, 'The politics of talk: Coming to terms with the 'new' scientific governance', *Social Studies of Science*, vol. 36, no. 2, pp. 299–320.

Jasanoff, S 1990, *The Fifth Branch: Science Advisers as Policymakers*, Harvard University Press, Cambridge, MA.

Jasanoff, S 2003c, 'Breaking the waves in science studies: Comment on H.M. Collins and Robert Evans, 'The third wave of science studies''', *Social Studies of Science*, vol. 33, no. 3, pp. 389–400.

Jasanoff, S 2005, 'Judgement under Siege: The Three-Body Problem of Expert Legitimacy', in Maasen, S and Weingart, P (eds), *Democratization of Expertise? Exploring Novel Forms of Scientific Advice in Political Decision-Making*, Springer, Dordrecht, the Netherlands.

Kahan, DM, Peters, E, Witlin, M, Slovic, P, Larrimore Ouelette, L, Braman, D and Mandel, G 2012, 'The polarizing impact of science literacy and numeracy on perceived climate change risks', *Nature Climate Change*, vol. 2, pp. 732–735.

Maasen, S and Weingart, P 2005, 'What's New in Scientific Advice to Politics?', in Maasen, S and Weingart, P (eds), *Democratization of Expertise? Exploring Novel Forms of Scientific Advice in Political Decision-Making*, Springer, Dordrecht, the Netherlands.

McNie, EC 2007, 'Reconciling the supply of scientific information with user demands: An analysis of the problem and review of the literature', *Environmental Science & Policy*, vol. 10, no. 2, pp. 17–38.

Porter, J and Dessai, S 2016, *Is Co-Producing Science for Adaptation Decision-Making a Risk Worth Taking?*, Centre for Economics and Climate Change Policy, Working Paper No. 263, Sustainability Research Institute, Paper No. 96, March 2016, viewed 19 December 2016, www.cccep.ac.uk/publication/is-co-producing-science-for-adaptation-decision-making-a-risk-worth-taking/.

Sarkki, S, Tinch, R, Niemala, J, Heink, U, Waylen, K, Timaeus, J, Young, J, Watt, A, Nezhover, C, van den Hove, S 2015, 'Adding "iterativity" to the credibility, relevance, legitimacy: A novel scheme to highlight dynamic aspects of science-policy interfaces', *Environmental Science & Policy*, vol. 54, pp. 505–512.

Sarewitz, D 2004, 'How science makes environmental controversies worse', *Environmental Science & Policy*, vol. 7, no. 5, pp. 385–403.

Skeptical Science 2015, *Media Coverage of The Consensus Project*, Skeptical Science, viewed 19 March 2015, www.skepticalscience.com/republishers.php?a=tcpmedia/.

Tang, S and Dessai, S 2012, 'Usable science? The UK climate projections 2009 and decision support for adaptation planning', *Weather, Climate and Society*, vol. 4, no. 4, pp. 300–313.

Tangney, P and Howes, M 2016, 'The politics of evidence-based policy: A comparative analysis of climate adaptation in Australia and the UK', *Environment and Planning C: Government and Policy*, vol. 34, no. 6, pp. 1115–1134.

Weichselgartner, J and Kasperson, R 2010, 'Barriers in the science-policy-practice interface: Toward a knowledge-action-system in global environmental change research', *Global Environmental Change* vol. 20, no. 2, pp. 266–277.

Weinberg, AM 1972, 'Science and trans-science', *Minerva*, vol. 10, no. 2, pp. 209–222.

Weingart, P 1999, 'Scientific expertise and political accountability: Paradoxes of science in politics', *Science and Public Policy*, vol. 26, no. 3, pp. 151–161.

Wilbanks, TJ and Kates, RW 1999, 'Global change in local places: How scale matters', *Climatic Change*, vol. 43, pp. 601–628.

6 Perceptions of the usefulness and usability of climate science and evidence for policy

That which can be asserted without evidence, can be dismissed without evidence.
Christopher Hitchens

In this chapter I present research gathered from 34 interviews[1] conducted in the UK and Australia with policy players involved in climate adaptation decision-making, to reveal their perceptions of the usefulness and usability of climate science and policy evidence. Interview participants were divided between *Climate Change Scientists* involved in the production of climate science outputs for government; *Policy Scientists* involved in the development of adaptation science and policy evidence such as impact and risk assessments that are often derived from climate science; and *Policy Players* working either in ministerial departments, non-departmental government agencies, or in or with local government councils who have used (or have been expected to use) climate-related evidence for policymaking and implementation.

My analysis shows how, given the nature of the problems it seeks to inform (see Chapter 4), norms, values and politics are important constituents in the development, use and perception of effective evidence for adaptation policy. Where evidence is perceived to lack usefulness and usability, conflict also often exists between the values and ideals implicit or inscribed within that evidence and the implied values of policy players, or within the policymaking norms and rules by which they abide. Investigating policy players' perceptions of evidence, therefore, is a useful means of elaborating on the tensions that exist between expert and political authority at the science-policy interface.

The views of UK and Australian policy players is not presented here as a direct comparative analysis between case-study regions. For simplicity and to avoid excessive compartmentalisation, interview results are presented in relation to my revised knowledge systems framework described in Chapter 5. Even so, important comparisons between the cases of Queensland and the UK presented here highlight the importance of reconciling the credibility, legitimacy and salience of evidence for climate adaptation policymaking. As discussed in previous chapters, there is no simple or linear interaction between science and policy for climate change policymaking, as per the linear-technocratic schema. As climate science approaches the spheres of policymaking and politics, it is manipulated in ways that ensure it is comprehensible and acceptable to policymakers. Thus, experts must address normative/political questions in the development of their research outputs, just as other policy players provide important subjective and normative input to expert evidence to make it useful and usable. Here I investigate policy players' perceptions of two separate outputs from the evidence-based policymaking process (see Chapter 4, and Figure 6.1 below).

First, I examine the use of the available *climate science*, comprising projections and scenarios that describe potential changes in climate variables at global, regional and even local scales. My research confirms the findings of previous analyses that climate science lacks salience for policymaking purposes in both the UK and Australia (Rickards et al., 2014; Tang and Dessai, 2012). My analysis, however, also reflects an increasing awareness in recent years by policy players of the nature and extent of the uncertainties associated with the science of *future* climate change; the need for other forms of evidence as a priority for effective adaptation policy; and, the difficulties of using climate science in the ways expected by linear-technocratic policy models.

Second, I examine policy players' perceptions of *policy evidence*: the *ex ante* risk and impact assessments that use climate and adaptation science and which seek to inform government decision-making about the potential implications of climate change for government policy. This analysis highlights the contingent character of policy evidence as distinct from the contingencies associated with the climate and adaptation science from which it is often derived, thus helping the reader to understand the extent of the overlap that exists between expert and political authority for policymaking. Climate impact and risk assessments, it seems, are often deficient across all three criteria of salience, credibility and legitimacy for many stakeholders involved in adaptation policymaking, while

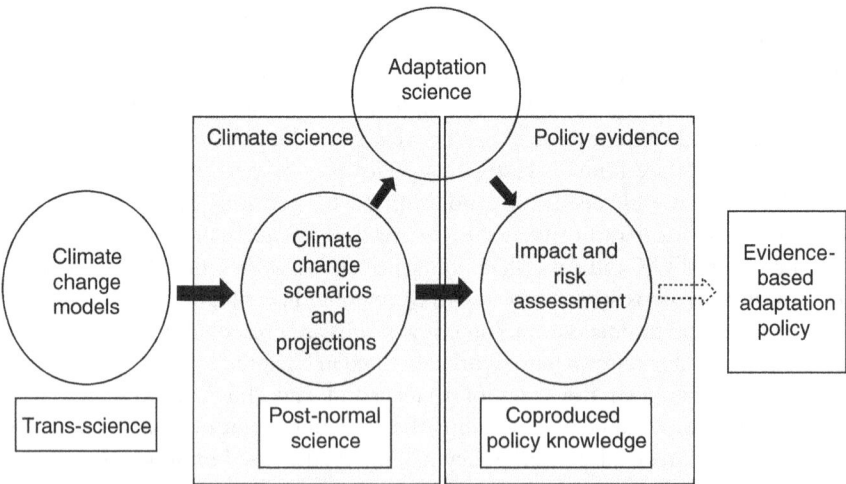

Figure 6.1 Climate science versus policy evidence: A two-stage analysis

nonetheless being used to bolster the legitimacy of government decisions on the basis of impartial expert authority. Interview participants describe how policy evidence has been subject to covert politicisation facilitated by the linear-technocratic assumptions of these assessment methods. I argue that how policy players perceive the effectiveness of policy evidence also reflects the difficulties of objectively understanding adaptation problems for the purposes of cross-level governance.

The various modes by which climate science and policy evidence may be perceived as lacking credibility, salience and legitimacy are summarised in Figure 6.2.

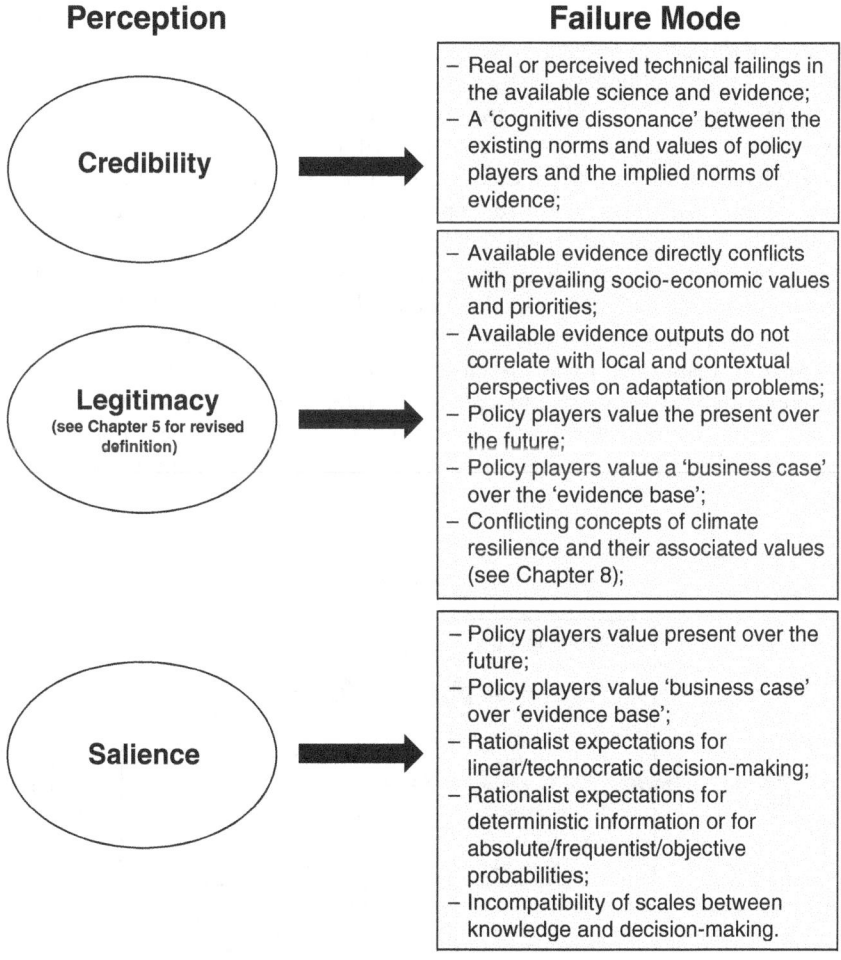

Perception	**Failure Mode**
Credibility	– Real or perceived technical failings in the available science and evidence; – A 'cognitive dissonance' between the existing norms and values of policy players and the implied norms of evidence;
Legitimacy (see Chapter 5 for revised definition)	– Available evidence directly conflicts with prevailing socio-economic values and priorities; – Available evidence outputs do not correlate with local and contextual perspectives on adaptation problems; – Policy players value the present over the future; – Policy players value a 'business case' over the 'evidence base'; – Conflicting concepts of climate resilience and their associated values (see Chapter 8);
Salience	– Policy players value present over the future; – Policy players value 'business case' over 'evidence base'; – Rationalist expectations for linear/technocratic decision-making; – Rationalist expectations for deterministic information or for absolute/frequentist/objective probabilities; – Incompatibility of scales between knowledge and decision-making.

Figure 6.2 How credibility, legitimacy and salience may be compromised by the norms and values of policy players

How useful are climate change projections and scenarios?

In the following section,[2] I investigate perceptions of climate change projections and scenarios (see Figure 6.3).

As discussed in Chapter 3, successive UK governments have invested considerable resources in the development of a portfolio of usable climate science to inform decision-makers. The UK-Climate Impacts Programme's (UKCIP) scientific products align with an ongoing bipartisan commitment (at least in theory) to evidence-based policy, an expectation that climate science should be used for statutory reporting requirements mandated under the Climate Change Act (2008), as well as for informing policymaking and implementation more generally (DEFRA, 2009; UK-Policy Scientist 1; UK-Policy Players 1, 4). In the Queensland case, by contrast, there has been much less focus on the production of standardised climate science outputs for policymaking, perhaps as a result of the lack of political consensus concerning the need for climate change policy. Queensland policy players have nonetheless relied on the climate science outputs of the IPCC, CSIRO, the Australian Bureau of Meteorology, and by external consultants, amongst others (Rickards et al., 2014).

While the attributes of credibility, legitimacy and salience in policymaking may be interdependent, these concepts can also be difficult to differentiate for policy players (Tang and Dessai, 2012; Weichselgartner and Kasperson, 2010). In the course of my research, participants' views of the credibility of the science were sometimes tied up with ideas about the degree to which that science adequately incorporated contextual viewpoints, was developed in a fair and transparent manner and was therefore considered legitimate. My analysis will differentiate between these concepts where interviewees made clear they were referring to one concept over another, but I also describe these concepts in tandem in circumstances where they were engaged simultaneously.

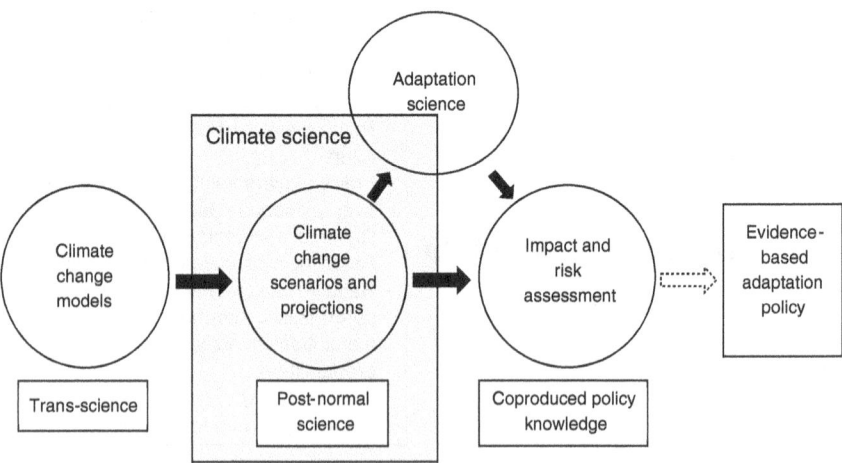

Figure 6.3 Part one: The credibility, salience and legitimacy of climate science

Climate science is credible, yet it has technical inadequacies

Although, when asked, interview participants almost universally agreed that climate science is credible, some subsequently appeared to contradict themselves, calling this credibility into question when pressed about their views about climate science's usefulness and usability. For instance, some suggested that some attempts to make climate science more salient had weakened its credibility (Aus-Policy Scientists 1, 2; UK-Policy Scientists 3, 10). As one UK-Policy Player commented in relation to UKCP09's probabilistic climate change projections:

> I think [UKCP09] tried to answer what the practitioners said they wanted but … in no other field do you try to quantify the unquantifiable. To a certain extent the scientists, i.e. the UKCIP fraternity, tried to provide the answer even though the question may not have been answerable.
>
> (UK-Policy Scientist 10)

Some openly questioned the credibility of climate change projections as a determinant of pre-emptive policymaking in the ways that adaptation policymaking had attempted to use them to date:[3]

> My opinion is that the projections are nowhere near good enough … as a modeller, I believe in climate change, I believe something is happening, but I don't believe the models … it's a really complex thing to cover, and to try and build all the micro-physics based rules to projections in the long term to say we definitely know what's going to happen in 50 years' time.
>
> (Aus-Policy Scientist 1)

> I think there's also a growing recognition that the very traditional top-down way of doing climate projections is not necessarily the most appropriate way.
>
> (UK-Climate Change Scientist 5)

Others suggested that the credibility of climate change projections was maintained on the basis of the famously near-universal expert consensus on climate change, rather than on the strength of the available evidence per se (UK-Policy Scientists 9, 10). Moreover, common simplifications used by policy players in the UK concerning the conclusions of the available evidence, such as the oft-used mantra 'warmer wetter winters; hotter drier summers' (Carrington, 2014; The Met Office, 2009) often failed to adequately account for the strength of that evidence and the many uncertainties involved.

A number of Australian and UK participants expressed concern that climate science was unable to account for the full range of uncertainties associated with future climate change (Aus-Policy Scientist 1; UK-Policy Scientists 9, 10; UK-Policy Player 4). The outputs from UKCP09 in particular lacked credibility for some participants since its assessment of probability of future climate outcomes did not cover the full range of uncertainty nor provide discrete estimates of likelihood of occurrence (UK-Policy Scientists 7, 9, 10). Alternatively, some noted that the presentation of uncertainties could give a false impression of the reliability of projections:

> We often went down the route of looking at some of the finer resolution projections coming out of UKCP09 but the problem is that it seemed to then cause a false level of certainty in peoples' minds around what they actually know, and then when you took a step back and said well actually the error bars are so huge it could be plus or minus then in a way it invalidates that level of precision they were looking for.
>
> (UK-Policy Scientist 7)

Meanwhile, both Australian and UK participants felt that it was becoming increasingly difficult to maintain the credibility of climate change projections amongst non-expert policy players and the public given the lack of congruence between projections and the observational weather and climate record in these two locations over the past 15 years (Aus-Policy Scientist 2). In the UK, for example, although drier summers and warmer winters are projected by UKCP09, there has been a recent trend of wetter summers and a series of very cold winters[4] (UK-Climate Change Scientist 5).

Climate science is not legitimate when it conflicts with prevailing values and political priorities

In Australia, participants believed that scientific uncertainties had been used by policy players to fortify political positions that conflicted with the conclusions of experts (Aus-Policy Scientists 1, 2, 3):

> Stakeholders who have got a position, select [interpretations of] science to suit their position, not to make a decision.
>
> (Aus-Policy Scientist 1)

Interviewees advising and working in government departments at various levels indicated that the science of climate change is often insufficiently legitimate to alter conflicting political or economic priorities or to meaningfully inform or direct relevant policy issues (Aus-Climate Change Scientists 2, 6; Aus-Policy Scientists 4, 6; Aus-Policy Player 1, 2):

> Whether we have political bipartisanship or partisanship affects how people view climate science ... people will believe something because the politicians are saying [it] rather than because of their understanding of climate science.
>
> (Aus-Climate Change Scientist 2)

This does not mean that climate adaptation does not occur in Australia under the guise of concurrent policy portfolios. As one prominent climate scientist noted in interview, what this lack of legitimacy for climate science means in practice is that conservative Australian governments at all levels have continued to pursue climate adaptation without mentioning climate change (Aus-Climate Change Scientist 6). Alternatively, adaptation is enhanced through concurrent policy priorities such as disaster risk management (Forino et al., 2016).

Interviewees suggested that policy players in Australia often do not question that the climate is changing, or even that humans have an influence upon it, but may use the associated uncertainties to question the legitimacy of scientific conclusions concerning the extent of human influence, thereby legitimising contrasting political positions (Aus-Climate Change Scientists 2, 6):

> Once [climate change] enters the political realm, people seem to be able to turn off the rational part of their brain that says … the science is telling me this and I should believe it, and they flip to 'I follow that political party or that line of [political rhetoric]' and for whatever reason that seems to over-ride any rational understanding.
>
> (Aus-Policy Scientist 2)

Thus, climate science has held limited legitimacy for policymaking in Queensland in recent years. As I argue in Chapter 3, these views reflect an ongoing conflict between the values underlying climate science and evidence and those policy priorities of successive local, state and federal governments. This conflict of values has also resulted in policymakers in state government passing adaptation issues to local government where issues of instrumental salience ensure that climate science is unlikely to be used:

> The political system is pushing the decision-making on how to implement the evidence to the lowest level [of government], a level that has the least capacity and ability to be able to understand and apply it … the planning scheme … probably doesn't give them the ability to consider climate change and what its true impact is.
>
> (Aus-Policy Player 2)

These findings suggest that evidence legitimacy is determined principally by political values and preferences rather than by any objective rationality provided by expert authority.

In the UK, a similar lack of legitimacy became manifest in the Cameron government's position on climate change. Although adaptation policy is required from central government and amongst government regulators mandated under the Climate Change Act (2008), the requirements placed on local government for reporting adaptation policy development and implementation were scrapped in 2010:

> If there's different messaging from the very top, again it gets picked up by
> the people we might be wanting to work with, and it does make it harder
> in some respects to win the argument that they need to do things differently.
>
> (UK–Policy Scientist 8)

This shift reflects a relative downgrading of adaptation and climate change
policy more generally by the Cameron government from 2010 to 2016
(UK–Policy Scientist 5; UK–Policy Players 1, 2, 4) (see also Chapter 3).

In the UK a concurrent perceived lack of salience for the available science
(discussed below) appears to have been used to delay policy action in a way that
indicates an increasing lack of legitimacy for climate science. Rather than
helping people to understand the nature of climate change and the associated
uncertainties, and therefore to address adaptation problems more effectively,
climate science information may actually foment the propensity for inaction on
the basis that people don't know how to deal with the uncertainties and so
decide to do nothing, or simply keep asking for more and better research
(UK–Policy Player 4; UK–Policy Scientist 4, 5). As discussed in Chapter 3, this
tendency toward inaction on the basis of inadequate evidence is reflected in the
content of many existing adaptation plans in the UK, as well as the various (now
defunct) climate change adaptation plans in Queensland, whose principal focus
had been on uncertainty reduction, rather than taking decisive measures in the
face of that uncertainty (see for example, EA, 2010a; Dedekorkut et al., 2010).

This apparent close correlation between the salience and legitimacy of
climate science is supported by the ongoing conflict between the types of
decisions expected during climate change adaptation policymaking and the
underlying norms of government relating to their preferences for short-term
policy priorities and incremental change. In the UK, although local government
policy players may explicitly state a wish to pursue prescriptive climate change
policies and, superficially at least, believe climate change projections are
credible, this science is nonetheless insufficiently legitimate or salient[5] to justify
such action. The science cannot justify substantive policy action in the face of
government's neo-liberal priorities that preference short-term economic goals
and more immediate evidence relating to the costs and benefits of action or
inaction. As one UK interviewee put it:

> Evidence base is interesting, but in our organisation now, it's the business
> case, it's the, why should we be doing it? what's the economic benefit?
> And it's not good enough to say, oh you know, if you do it now it's
> cheaper than having to deal with it in ten years' time, [be]cause again that
> comes back to the robustness of the 'fact' that you are going to have to deal
> with it in ten years' time.
>
> (UK–Policy Player 2)

Similar views were expressed by a number of UK and Australian participants
working at the science–policy boundary. Just because the science suggests that

one ought to take action from a precautionary point of view does not mean that the evidence provides sufficient justification for action amidst a range of competing interests, priorities and sets of evidence. The conclusions drawn by climate adaptation scientists are based on a set of expert attitudes to climate risk (e.g., concerning the robustness of projections of future climate change) that are not necessarily shared by a wide range of stakeholders.

On this note, a prominent Australian policy scientist described how discussions with civil servants working in Commonwealth ministerial departments on the issue of climate change adaptation had confirmed how the simple presentation of evidence to demonstrate the need to adapt had been insufficient on its own to justify policy, even if the conclusions that could be drawn from that science appeared to unambiguously suggest obvious courses of action:

> What I've just described has arisen out of us taking evidence about climate change to things like the Department of Climate Change and saying 'we ought to be adapting to that' and them saying 'eh, why do you think that?' or 'that's not a good enough [justification]' Just saying there's an impact and even that there's an impact and we know what could be done to adapt is nothing like enough to make a convincing case for us to take to Treasury.
>
> (Aus-Policy Scientist 4)

Climate science lacks salience due to insufficient information about the vulnerability and sensitivity of receptors

The interpretation and use of climate change scenarios and projections has clearly been a difficult process for decision-makers concerned with climate adaptation policy, and not just in terms of a lack of political acceptability. There was general agreement amongst those interviewed in both the UK and Australia – confirming previous studies in both countries (Tang and Dessai, 2012; Rickards et al., 2014) – that these scientific outputs lack salience for policymaking. This, despite a persistent popular expectation that such evidence is, and should be, a principal source of information useful for effective climate change policymaking (see for example, Church et al., 2016; Pitman, 2016).

Although the science of future climate change is important, most interview participants agreed that other forms of evidence are also required for policymaking:

> Climate science … pretty well has done all it can do. And you really have to start at the vulnerability perspective and really understand the vulnerability, at which time climate is a source of information that has to be considered.
>
> (UK-Policy Scientist 5)

Contrary to popular perceptions, projections of future climate change are often a secondary concern given the uncertainties associated with it (see Chapter 4).

Indeed, evidence relating to future climate change was considered to be only truly meaningful in the context of adequate understandings of vulnerability and exposure to climate for any individual risk-receptor, community or government jurisdiction. As one UK interview participant noted of UKCP09 in relation to its use for local government decision-making:

> It's like handing somebody a dictionary, you can start reading the dictionary at page 1 and you'll learn stuff, but it only becomes useful to you when you actually know what you're looking for, and I think with UKCP09 you have to have done a little bit of groundwork to understand what … kinds of impacts you've already seen, if you're lucky what kind of thresholds you've already seen.
>
> (UK-Policy Player 1)

This sort of insight was generally provided by those with direct experience of using climate science for risk/impact assessments and the development of adaptation plans and policy, rather than by those who advocated its use but had no real experience of using it for policymaking in practice (UK-Policy Players 1, 2, 3, 4; UK-Policy Scientists 5, 6, 7, 8, 9; Aus-Policy Scientists 4, 5; Aus-Climate Change Scientist 1):

> It was never the case that [for] any adaptation decision you have to just go straight to the projections and they'll give you the answer, actually it's much more complicated than that, and in quite a lot of cases you don't need to use them at all.
>
> (UK-Policy Player 4)

A majority of those interviewed across both case-studies believed that the most important form of evidence required for effective adaptation policy and decision-making is an improved understanding of community and organisational vulnerability and exposure to existing climate variability, and to a lesser extent, their vulnerability to future climate change:

> To understand the climate change impacts first you have to understand this situation with or without climate change in some respects.
>
> (Aus-Policy Scientist 6)

These perspectives contradict long held expectations within both climate science and policymaking circles that climate change projections are the principal source of evidence required for adaptation policy or that, on its own, climate science can inform decision-making in a direct and instrumental way (UK-Policy Scientists 5, 9; Aus-Policy Player 1; UK-Climate Change Scientists 4, 5; Dessai and Hulme, 2004; Church et al., 2016; Pitman, 2016).

Evidence relating to extant climate exposure, vulnerability and resilience, however, are often missing when making (in particular, strategic) adaptation decisions. In order to fill this gap a range of expert and non-expert policy players should be involved in the development of evidence. Some believed that local authorities, and the businesses and communities they serve, constitute a largely untapped information resource for understanding potential climate change impacts. These stakeholders had access to valuable data about weather-related damage to both public and private infrastructure and the costs of its repair. If collated appropriately and used in conjunction with the available climate science, this data could be used effectively to understand the risks from climate-related extreme events (UK-Policy Player 1; Aus-Policy Player 5). Unfortunately, this type of information is often, as yet, unavailable; or systems that have been put in place allowing this data to be collected for the purposes of national adaptation policy are as yet under-developed (UK-Policy Players 2, 3). Thus, the difficulties of applying science to policymaking remain, whereby, effective adaptation often requires location and context-specific knowledge that is either unavailable, difficult to use for the purposes of strategic policymaking due to its lack of generalisability, and/or too costly to collate (Adger, 2011; Cash et al., 2006).

Competing types of evidence are more salient than climate science at relevant temporal and geographic scales

It seems clear, from the perspectives described above, that even where the evidence-based mandate is strongly adhered to for adaptation policymaking, climate science struggles to provide the types of relevant information decision-makers seek. In the UK, although enshrined in the reporting requirements of the Climate Change Act (2008), this mandate is interpreted differently at local government level, where policy players have increasingly fewer obligations to use climate change projections due to their lack of salience and legitimacy. Nonetheless, useful, usable evidence for local government policy is increasingly required for the development of a 'business-case' rather than simply an evidence base.

In order to justify efforts toward climate adaptation, therefore, policy players must provide decision-makers with an evidence-based assessment of the costs and benefits of action derived from what is actually known about socio-economic and ecological systems and for which science and engineering can provide robust estimates over the short-to-medium term. In both countries, such fiscal evidence aligns with the decision-making norms and priorities of contemporary government relating to neo-liberal ideals and short-term economic concerns and sits in contrast to the available climate science which generally has a longer-term focus and is necessarily more uncertain. The latter, more often than not, fails to be suitably relevant on its own amongst competing economic evidence that is considered more appropriate for use in economic-rationalist cost-benefit assessment (UK-Policy Players 1, 2, 3; Aus-Policy Player 5).

Climate science lacks salience because it is complex, uncertain and prone to misinterpretation

Irrespective of the availability of evidence relating to climate vulnerability and exposure that would make climate change projections more usable, or the availability of climate science that would be relevant at appropriate scales, there was almost universal consensus from both Australian and UK participants that many existing scientific outputs describing future climate change hazards are often simply too complex and uncertain to aid decision-making. In the UK, participants complained that the information provided by UKCP09 is difficult to access and to understand; the key messages difficult to discern and prone to misinterpretation or over-simplification; that they provide abstract information about climatic variables that does not adequately relate to decision-making needs, or the inability of this evidence to provide information about the most pertinent climatic or weather-related variables. As one interviewee stated in relation to UKCP09:

> Very few people use the level of detail it has gone into and for most people they're completely bamboozled by it, including not just the user interface side of things ... but also the projections document ... [and] the sort of high-level findings ... [be]cause a lot of people, the majority of people going to it are actually looking for something quite simplistic ... they maybe just want one or two maps and they're not that easy to find ... and a lot of the way it's written is from a very technical perspective telling you what you can't do and what's inappropriate.
>
> (UK-Policy Player 7)

Climate projections provide data that is often just too abstract and that does not provide an adequate sense of how the weather or climate is going to change in terms of the frequency of extreme events or even what any particular change in average temperatures will actually be like. For this reason many believed that there was too much emphasis placed on climate projections as a tool for informing adaptation, when much of the time it lacked practical instrumental relevance for decision-making (UK-Policy Scientists 6, 7; UK-Policy Players 1, 7; UK-Climate Change Scientist 5):

> You can be over dependent on projections, I think they're useful for driving a range of scenarios ... but be careful in not over-interpreting them, even if you are working in the scientific field.
>
> (UK-Policy Scientist 7)

In the UK these types of difficulties were attributed to the use of probabilistic projections, which are a dominant feature of UKCP09, often presented in the form of probability density functions (PDFs) and cumulative distribution functions (CDFs) (see Figure 6.4 below).

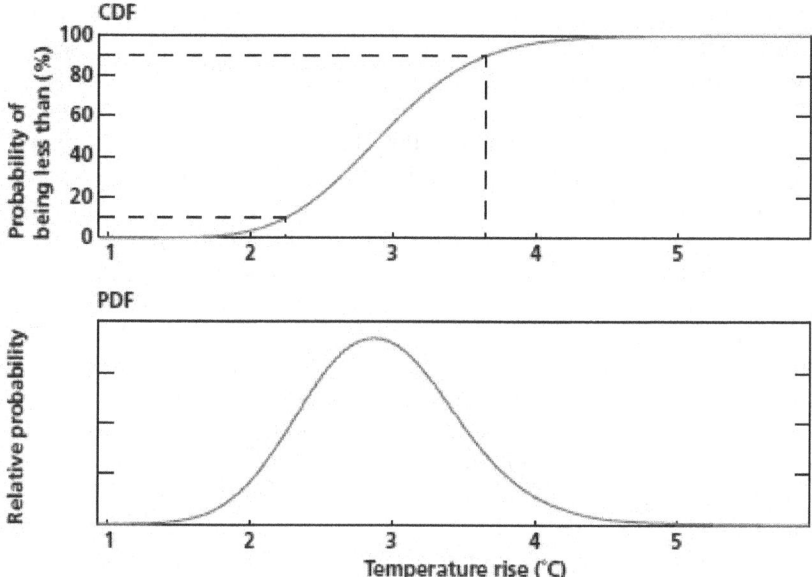

Figure 6.4 Top panel: Cumulative distribution function (CDF) of temperature change for a hypothetical choice of emission scenario, location, time period and month. Bottom panel: The corresponding probability density function (PDF) for this hypothetical case (adapted from Murphy et al., 2009: p. 24)

PDFs and CDFs attempt to provide a probabilistic estimate of the range of possible future climatic states. These outputs indicate the level of congruence of potential outcomes with estimates of future climate change provided by an ensemble of general circulation models (GCMs). Such *Bayesian*[6] probability distributions are not equivalent to conventional frequentist or other empirically derived probabilities of occurrence and cannot be used to provide an objective assessment of the relative probability of a particular climatic outcome occurring; rather, they attempt to indicate the range of likely uncertainty based on our current understandings of anthropogenic forcing of the climate system (Frigg et al., 2013b). As such, although UKCP09 expresses climate change hazards via PDF and CDF outputs for climatic variables such as mean annual temperature, precipitation and humidity (Murphy et al., 2009), UKCIP warns against using specific probabilities from these outputs when developing adaptation plans or understanding impacts. Most non-expert policy players interviewed (along with some experts) who had attempted to use the outputs of UKCP09 agreed that it took some time to understand how to access the information available through the online user interface, to understand the nature of the outputs and how they differed from conventional probabilities, and that for many decision-making applications they lacked usability. 'I can't imagine how an ordinary user sort of copes really … it is horrendously complicated I have to say' (UK-Climate Change Scientist 5).

Notwithstanding the lack of salience of such outputs for policy decision-making, many expert participants believed that UKCP09 had been useful, or may yet prove to be useful for technical modelling applications seeking to understand climate impacts, for example, in relation to flood risk and water resources modelling. Further, UKCP09 had been useful for helping to understand the nature of the uncertainties associated with climate change. This testimony supports the work of Weiss (1979) that evidence can fulfil multiple roles besides directly (or instrumentally) informing or directing policy decisions. In the case of climate science, though it may not be wholly salient for the purposes of *problem-solving*, this science can nonetheless fulfil roles in ways that align with *enlightenment, tactical* or *political* modes of evidence use (Weiss, 1979).

In Australia, participants agreed that climate science does not provide salient information for decision-makers. By contrast however, the climate science outputs available in Australia are less well developed and are presented in a more deterministic format, providing climate change scenarios for a limited number of specific timescales, emissions trajectories and GCMs at various points to 2100 (e.g., CSIRO, 2016). The reasons given for this perceived lack of salience related most commonly to the spatial (global and regional) and temporal (multi-decadal) levels at which this climate science seeks to describe future hazards, rather than due to the difficulties of accessing and interpreting the available information. The scales and levels at which this information is presented are problematic since they do not generally correlate to the (principally local) governance levels and short (decadal) timescales over which adaptation policies are developed and implemented (Aus-Policy Players 1, 2, 5; Aus-Policy Scientists 3, 6). One policy player in Australia stated that she found some of the messages derived from climate change projections to be:

> meaningless, when it's less than a degree centigrade … no one understands that … and sea-level rising of you know 400mm or 600mm … 400mm that's nothing. No one really had the understanding of the enormity of that incredibly small change across the huge surface of the Earth and what that really meant.
>
> (Aus-Policy Player 1)

Climate science is poorly communicated

A compounding issue in providing salient, legitimate climate science was the tendency for poor communication and dissemination of climate change projections by the scientific community (Aus-Policy Scientists 1, 2; UK-Policy Scientists 5, 6, 7, 8, 9, 10). Producers of evidence such as UKCP09 had not adequately communicated what these products were and how they should be interpreted. In part this difficulty originated due to the nature of the uncertainties and the caveats that scientists placed on their conclusions (described above) which made it difficult to discern clear messages. As one interviewee described UKCP09:

the idea that it was probabilistic ... the way they showed the results ... people exhibiting the results gave caution ... about trying to ignore the central tendency and then they proceeded to show us loads of pictures that were driven by the central tendencies.

(UK–Policy Scientist 9)

Another participant described the communication and dissemination of UKCP09 as follows:

There was such a preciousness about not saying the wrong thing that actually nobody quite said the right thing ... there was a whole use of language and I struggle even now to try and reproduce it but it's something [like]: it's highly unlikely to be less than X degrees or whatever, which just doesn't trip off the tongue you know, and actually somehow you've got to summon up the courage to ... suggest the mean, or the average is around 2 degrees of warming ... and I think it was a real communication barrier around CP09, the use of language.

(UK–Policy Scientist 6)

An important contrast between the case-studies was that the emphasis placed on the salience (or lack thereof) of the available climate science for adaptation decision-making was considerably less amongst Australian participants than amongst their UK counterparts. This can be explained by the contrasting political contexts of the two cases, described in detail in Chapter 3. In particular, the lack of bipartisan political consensus on the issue of climate change in Australia may have meant that interview participants considering the use of climate science for policy were generally more concerned about the political acceptability of pursuing climate change adaptation policy, or the legitimacy of climate science for decision-making in concurrent related policy fields such as disaster risk management, urban planning and water resources management. As such, Australian policy players' priorities were less focused on the practical usability or salience of climate change projections than they were on the political ramifications of using them in the first place. This finding points to fundamental issues relating to the legitimacy of climate science as the priority focus for Australian policymaking. Given the climate extremes experienced in Australia and the ongoing focus of public policy on disaster risk management and water resource management as a result (Heazle et al., 2013), interviewees alluded to numerous opportunities for incorporating climate science into the development of existing policy issues but for which there was insufficient legitimacy in that science for its use to be deemed politically acceptable (Aus-Policy Players 1, 2, 3; Aus-Climate Change Scientist 6; Aus-Policy Scientist 4).

How useful are climate impact and risk assessments?

Impact and risk assessment methods have been used to varying degrees and with varying formats in Australia and the UK for many years. In particular, they are used for the purposes of strategic public policy, for urban and regional planning schemes, for environmental management such as wildlife conservation, and in the provision of public infrastructure such as flood defences and water resource facilities. A number of the issues described here mirror the difficulties encountered in relation to the usefulness and usability of climate science described above. The following section,[7] however, highlights the extent of normative influence upon the development of policy evidence through *ex ante* impact and risk assessment for policy (see Figure 6.5), and illuminates the significant overlap that exists between expert and political authority in climate adaptation policymaking when using such methods.

This section also highlights an important distinction between the development of climate change adaptation policy in Australia and the UK. Whereas both political and evidence legitimacy is lacking (and therefore a priority concern) in Australia, in the UK, where political acceptability for climate change policy and evidence has largely been established under the Climate Change Act (2008), participants demonstrate greater concern for the technical adequacy (credibility and salience) of *ex ante* policy evidence. The imbalance of testimony presented below reflects this important difference between Australia and the UK in the contrasting focus given by interview participants upon legitimacy versus credibility and salience, depending on their priorities. This contrast is an important indicator for understanding one of the central arguments of this book, namely, that legitimacy holds primacy over credibility and salience in the development of adaptation policy evidence.

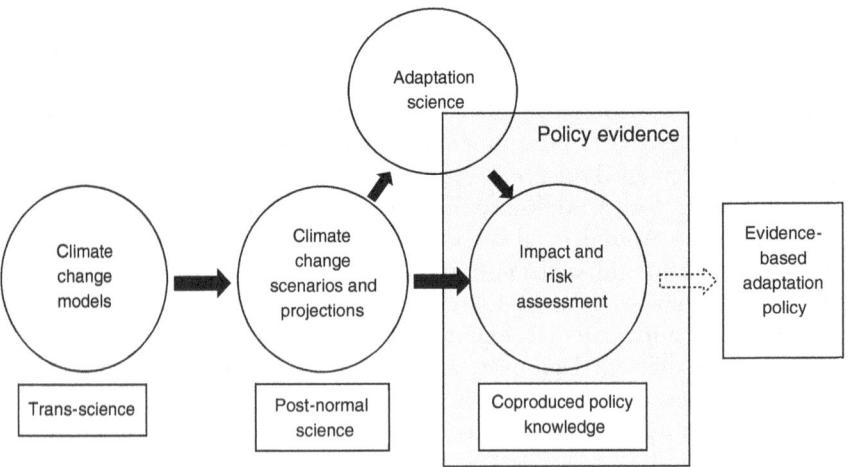

Figure 6.5 Part two: The credibility, salience and legitimacy of policy evidence

Ex ante policy appraisal cannot account for the complexity, nuance and uncertainties of climate adaptation

Due to the contingent nature of adaptation problems (described in Chapter 4), *ex ante* policy assessment necessitates normative and subjective input in the framing, interpretation and presentation of evidence (Owens et al., 2004). Even when overseen and/or conducted solely by scientific experts, I argue, these assessments are conducive to accusations of insufficient credibility as a source of privileged expert authority.

A number of interview participants expressed the view that risk assessments that follow a linear model of evidence provision for policy lack credibility because they cannot adequately account for the 'wicked' characteristics of climate adaptation and are therefore insufficient for informing adaptation policy. The interpretation of science through *ex ante* policy assessment does not provide wholly credible answers in line with the expectations of policy players and the public about how expert authority should inform policy (UK-Policy Scientists 5, 6, 7, 8, 9, 10; Aus-Policy Scientists 1, 2, 3, 4, 5; Aus-Policy Players 2, 4). For instance, participants in the UK-Climate Change Risk Assessment (UKCCRA) 2012 recounted how its credibility had been limited by the pursuit of traditional (linear-rationalist) risk assessment approaches, and by attempts to objectively and definitively apprehend adaptation problems through science and economics. Risk-based approaches that attempt to assess climate risk through empirical quantitative or semi-quantitative estimates of likelihood and consequence are inherently problematic in this regard (UK-Policy Scientists 1, 4; UK-Policy Players 1, 4):

> Well, you can't do that with climate change ... we don't look at it in terms of probability times [consequence anymore], you know, you have to take a much more qualitative approach to it.
>
> (UK-Policy Player 4)

Attempts to follow such a positivist approach during the derivation of the UKCCRA 2012 had met with a number of obstacles.

Most notably, the panel of experts who had assessed climate change risks relating to biodiversity and ecosystem services for the UKCCRA 2012 had rejected the method proposed by those technical contractors who were managing the assessment, on the basis that it could not account for the degree of complexity and uncertainty inherent in interdependent social-ecological systems (UK-Policy Scientists 1, 2, 3, 7). As a result, the assessment relied on a mutually agreed narrative describing the likely impacts from climate change. As one policy scientist noted, the assessment for this particular sector was 'qualitative and a far more moderated "opinion" based on the evidence rather than an attempt to follow the methodology too closely' (UK-Policy Scientist 3).

Despite abandoning the linear prescriptions of the UKCCRA 2012's method for this sector, the final assessment for biodiversity and ecosystem services was nonetheless also considered to lack credibility and salience on the basis that, in order to reach consensus between experts (thus ensuring the knowledge production process was seen as fair and unbiased, that is, legitimate), climate change impacts were described at such a high level of abstraction that the assessment ultimately failed to provide a technically credible (and salient) assessment that could benefit policy players' understanding of the risks and impacts of climate change (UK-Policy Scientists 2, 3, 6, 7, 8, 9). As one participant noted about trying to adequately understand climate change impacts:

> So you know ... you can get at it from an individual scientific point of view if you're a specialist, but actually putting that into some kind of generic analysis process, we're a way off that.
>
> (UK-Policy Scientist 2)

A number of UK participants spoke of the uncertainties associated with understanding the complex dynamics of social-ecological systems and then combining climate science with that knowledge to understand potential impacts and risks in a definitive way:

> We're dealing with a very complicated system in the natural environment and often the uncertainties in the ecological responses to climate change are at least as great if not greater than the uncertainties around the projections themselves.
>
> (UK-Policy Scientist 7)

Those with expertise in biodiversity and species conservation expressed the view that the risk assessment method used for the UKCCRA 2012 (see Chapter 7 for further details) had followed an engineering approach that was ill-suited to understanding the risks to social-ecological systems (UK-Policy Scientists 6, 7, 8, 9):

> It was an approach that was used but was rather forced somehow, I think it was made to work but it was a little stiff, and I think the contractors struggled with that.
>
> (UK-Policy Scientist 7)

These difficulties and the associated lack of credibility in the UKCCRA 2012, however, were not confined to adaptation problems with such high complexity and uncertainty as for ecosystem services. As one policy player involved in the UKCCRA 2012's expert workshop on the built environment noted:

> It very rapidly descended into people saying, 'you're missing the point here, everything's cross cutting, you know you can't talk about these things in isolation' ... I don't want to say it but it was almost a tick box exercise.
>
> (UK-Policy Player 2)

These difficulties speak to the interconnectedness of adaptation problems (described in Chapter 4), making them difficult to definitively define or resolve. These difficulties also relate to the lack of salience of climate science for determining climate risk, which was perceived to be particularly acute for highly uncertain and complex adaptation problems relating to social-ecological management.

One prominent UK-Policy Scientist described how the applications for using climate science through impact and risk assessment may be quite different depending on the type of problem being examined and its associated uncertainties. In relation to water resources or flooding, this scientist believed, adaptation problems are described by a small set of affected parameters relating to the viability of infrastructure or the means of managing a resource which an assessor can put strict engineering, jurisdictional, geographic and temporal bounds around. These types of adaptation issues are more amenable to using climate science and economic projections to understand specific variables of vulnerability and resilience. By contrast, for problems of biodiversity, ecosystem services and species conservation, it is much more difficult to define and delimit individual adaptation problems. There are so many relevant variables that the principal focus is on the movement of a social-ecological system in a particular direction and how that can be affected by human and climatic influences. The most helpful types of climate science for this type of adaptation policy and planning, are indications of the general direction of climate change and broad indications of its potential magnitude:

> No matter how good the probabilities or scenarios or other climate information in the UKCP[09] or other projections, the real challenge is lack of knowledge about sensitivity of different elements of the natural environment and potential tipping points, so sometimes it doesn't really matter whether there's a 60 per cent or a 90 per cent chance that it will become 3 degrees hotter because you're often not quite sure at what point something becomes too hot, or too wet or too dry.
>
> (UK-Policy Scientist 7)

Those interviewed who had been involved in the UKCCRA's 'expert elicitation' workshops recounted how these knowledge production events, involving a range of scientific, technical and industry experts, had been compromised in terms of their credibility due to a tendency toward the selection and prioritisation of risks:

1   that already had a considerable evidence base available about them, rather than on their actual potential severity;
2   that were already the academic or professional focus of those experts involved; and/or,
3   about which there was already a legislative provision for their management, which meant that risks were more easily characterisable (UK-Policy Scientists 1, 2, 7, 9, 10; UK-Policy Player 7).

An Australian policy scientist suggested that the use of expert elicitation processes and confidence levels to prioritise climate risks were increasingly inadequate or irrelevant for this kind of assessment because they are based on an assumption of the adequacy of expert's past experience, alongside their knowledge of climate science. Experts' knowledge and experience of the natural environment cannot fully account for the chaotic nature of bio-geophysical and social-ecological systems in the face of anthropogenic climate change. Experts cannot be confident about any particular outcome under a non-stationary climate and the deep uncertainties associated with social-ecological systems (Aus-Policy Scientist 5).

Others suggested that there was insufficient cross-sectoral analysis of risks during the UKCCRA 2012. Each of the 11 sectors covered by the UKCCRA prioritised ten risks however there was no cross-comparison of those risks that would assess their relative strategic priority (UK-Policy Scientists 3, 10; UK-Policy Player 7). As one UK-Policy Scientist noted:

> I'm not sure they fully thought through the chain of risks … with the technique they were using there was a risk that you could lose half of Kent [to sea level rise] and still be worried about your pension if you lived in Margate [a town on the Kentish coast].
>
> (UK-Policy Scientist 10)

Some also criticised the assessment for failing to account for international influences, trans-boundary interactions or indirect and compound risks from multiple risk interactions (UK-Policy Scientist 10; UK-Policy Player 7).

Mirroring concerns about the salience of climate projections in the absence of other information, another common indictment of the UKCCRA's credibility was that it failed to adequately consider the resilience and adaptive capacity of ecosystems, communities, industry and government institutions (UK-Policy Scientists 6, 7, 8, 9, 10; UK-Policy Players 1, 7):

> We were just a bit worried there wasn't enough emphasis on the capacity that people would have, because being able to be resilient and your level of climate change risk isn't just about what falls from the sky, it's also about your ability to handle it.
>
> (UK-Policy Player 1)

These findings appear to validate the political analysis presented in Chapter 3, in which I argue that, although the UK has been a world leader in advancing climate change adaptation policymaking, it has relatively little experience of extreme climate events and as a result its consideration of resilience and adaptive capacity is under-developed compared to the case of Queensland.[8]

The interpretation of climate science through *ex ante* policy assessment, however, can also be problematic. The UKCCRA had only presented climate risks as interpreted through the scope of UKCP09. This latter scientific product, however, was limited to the suite of General Circulation Models (GCMs) used for those probabilistic projections. One prominent policy scientist believed, therefore, that the UKCCRA had not adequately accounted for *known unknowns* relating to climate change which, although mired in uncertainty, the scientific community still know enough about to understand the relative severity of their potential consequences. This policy scientist (a national expert in sea-level rise and flood risk on the urban environment) believed that UKCP09 had not adequately accounted for extremes of sea-level rise associated with non-linear climatic and geo-physical changes. He believed that a credible and legitimate[9] risk assessment would have addressed such risks irrespective of their likelihood as derived by necessarily uncertain Bayesian probabilities (see Frigg et al., 2015), due to the potentially catastrophic consequences that could arise from such changes. The UKCCRA appears to have been subject to political influence in this regard (UK-Policy Scientist 10) (see Chapter 7) and was alluded to by a number of others interviewed. Another prominent UK-Policy Scientist described the UKCCRA as 'The political presentation of science' (UK-Policy Scientist 5). Indeed, political influence was perceived by many in both the UK and Australia, calling into question the credibility of these assessments on the assumption that policy evidence should principally be the product of impartial analysis and privileged expertise (UK-Policy Scientists 2, 7, 10; Aus-Policy Player 2, 3; Aus-Policy Scientist 5, 6).

In Australia, there had been less emphasis placed on the production of policy evidence such as climate change risk assessments because, as described above and in Chapter 3, there has been more protracted political debate about the legitimacy (i.e., political acceptability) of climate science and less emphasis on its subsequent use for policymaking. The lack of political legitimacy for climate change in Australia has resulted in a preoccupation with the legitimacy of climate science and policy evidence by Australian interviewees, and as a result, far less concern for evidence salience or credibility. Conversely, an established legitimacy for climate change and climate science in the UK (notwithstanding the aforementioned difficulties at local government level) has resulted in a principal preoccupation with evidence salience.

This dearth of political and evidence legitimacy for climate change in Australia has also meant that, although many Australian participants had some experience of undertaking local and regional assessments of climate change impact and risk, they lacked experience with strategic or federal government-led policy evidence development, such as for the UKCCRA 2012. This meant that they were less familiar with the types of cross-scalar governance and political difficulties described by UK participants. Nonetheless, a number of Australian interview participants expressed concern that risk assessments lack credibility because they fail to account for the plurality of valid perspectives (Aus-Policy Scientists 1, 2, 5, 6; Aus-Policy Player 2). Echoing the views of UK participants, one former Queensland state government official noted how generic assessments fail to account for the nuance and complexity of adaptation issues:

> Every single catchment, every single bit of hydrology in our State is very very different ... so it's impossible to simply apply a broad brush [assessment] across the board.
>
> (Aus–Policy Player 2)

Risk assessments lack credibility because decision-making prescriptions are often misused and misinterpreted

Most participants agreed that, although it is a useful conceptual approach in theory, risk-based decision-making had been used in ways that are conducive to assuming that evidence can be used linearly, impartially and unproblematically to inform policymaking. *Ex ante* assessments, therefore, tend to be undertaken as a one-off assignment in a way that assumes an ability to apprehend the adaptation problem with sufficient accuracy and detail to allow policymakers to adequately understand the risks from climate change. As described above (and in detail in Chapter 4), however, such assessment is limited in practice due to the complex, uncertain and contentious nature of adaptation issues: 'That kind of simplistic process, which is a good start, is really wholly insufficient' (Aus–Policy Scientist 5). A prominent UK–Policy Scientist believed that many experts and policymakers still hold unrealistic and often naïve expectations about the role and use of evidence, assuming that the available evidence can inform policy in a linear, rational and comprehensive way (UK–Policy Scientist 5).

One possible signifier of these expectations was the fact that no one who worked for DEFRA on the 2012 UKCCRA was expected to work on its next iteration in 2017 (UK–Policy Scientist 5; UK–Policy Players 4, 5). A common working practice within UK ministerial departments is the continual turnover of staff across departments, whereby policy players are expected to change roles every couple of years (UK–Policy Scientist 5). Some interviewees believed that this practice would mean that newcomers to climate risk assessment would fail to learn from past *ex ante* assessments and, therefore, to understand the complex, contentious and pluralistic nature of adaptation problems and the difficulties of deriving and utilising evidence in pursuit of them. This situation would severely constrain opportunities for institutional learning, the development of adaptive capacity and thus, ultimately, the credibility and instrumental efficacy of adaptation evidence (UK–Policy Scientists 1, 5; UK–Policy Player 5). The 2012 UKCCRA, they believed, would ultimately be judged on the extent to which it was used as a first step in an iterative learning process (UK–Policy Scientists 3, 5; UK–Policy Player 4). However government appeared to have misused the assessment and misunderstood its utility in this regard.

This view was echoed by Australian participants. Climate risk assessments should be used as a means to prioritise risks within a 'socially engaged process' (Aus–Policy Scientists 3, 4, 5), and is problematic when used in a linear way assuming that once completed it will allow policy players to adequately understand risks and therefore resolve adaptation issues. Given the wicked characteristics of adaptation problems, interviewees stressed that the precision of any assessment is less important than how it is used within the overall policymaking process. A majority of assessments have been misused in this way as a result of rationalist expectations of evidence for policy (Aus-Policy Scientists 3, 4, 5). Although the literature often bemoans a lack of action on adaptation, this inaction actually results from unrealistic expectations that the expert community can and should provide answers that may not actually exist or be reliably answerable in advance of policy action (Aus-Policy Scientist 4).

Impact and risk assessments lack legitimacy when the values and interests of experts conflict with policymakers'

An underlying theme of much of the collected testimony for this book was that there is an ongoing mismatch of norms and values between different groups of policy players in the development of policy evidence. This mismatch was reflected in the general perception from those interviewed that *ex ante* assessments of evidence for policy necessarily lack legitimacy for informing policymaking decisions across varying scales and levels of governance. Nonetheless, most agreed that evidence coproduction processes are still necessary for adaptation policy, due to the nature of the uncertainties and complexities associated with adaptation issues, and as a result of the need for evidence beyond economic and scientific forms. The contingent character of risk assessment means that there will always be a limit to how legitimate it can be within the policymaking community since these assessments necessarily struggle to provide information in a way that is meaningful for all relevant policy players across varying levels of government (Cash et al., 2006).

While these assessments prioritise scientific, engineering and economic evidence, many believed that effective assessment also requires explicit subjective, normative choices about what is valuable, resilient and vulnerable to climate, and what level of climate risk is acceptable (UK-Policy Players 1, 2, 3; Aus-Policy Scientists 5, 6; Aus-Policy Players 1, 2; Climate Change Scientist 6). The political executive, however, is usually unwilling to mix explicit normative, political choices with technical assessment due to the perceived need to maintain the objectivity of legitimate expertise, even though such normative/political influence does inevitably and implicitly happen when addressing issues of vulnerability and resilience (Keller, 2009; Jaeger et al., 1998).

As described in detail in Chapter 7, the UKCCRA 2012 was ultimately considered to be a political document; it had been covertly influenced by the political executive in terms of the potency and tone of its messages because the initial technical assessment had not been politically acceptable to government departments. This politicisation suggests not only a lack of political acceptability of expert conclusions, but also suggests a lack of legitimacy in terms of the fair and unbiased nature of the knowledge production process (UK-Policy Scientists 1, 2, 3, 5, 7). In the case of the UKCCRA 2012, this politicisation appears to have been particularly acute since the assessment began under the reign of the

Labour Party and was completed under a Conservative/Liberal Democrat coalition:

> A change of politics does play a role in a document that's going to be laid before parliament ... the way that the information is presented is influenced by politics because it has to be signed off by the secretary of state.
>
> (UK-Policy Scientist 5)

The end result was that the UKCCRA's outputs were essentially trivialised and left unused by the Cameron government, in line with its downgrading of climate adaptation as a policy issue more generally whereby, since coming into power in 2010, it had scrapped most statutory requirements for adaptation that weren't legislated by the Climate Change Act (2008). Adaptation had been essentially downgraded to a 'tick-box exercise' to fulfil the evidence-based mandate. Most notably, national government had removed all obligations for local government and state infrastructure providers to report their climate change risks and adaptation actions.

Other participants in the UKCCRA confirmed this lack of legitimacy in terms of the political acceptability of the assessment's report. The de-prioritisation of adaptation policy was evident in the National Adaptation Programme unveiled in 2013 (DEFRA, 2013a) which, although as originally intended was supposed to be systematically informed by the UKCCRA, in the event had only tenuous links to this assessment. The new government had at that point changed their focus and prioritisation from direct intervention and management of adaptation issues, to acting as a facilitator to communities and private enterprise (UK-Policy Players 2, 5; UK-Policy Scientists 1, 5):

> The whole political climate changed from the point of commissioning the work [of the UKCCRA] to it being published and if there had been a different government in place at the time of publication that may have been more supportive of the climate change agenda, potentially more might have been made of it ... By the time we published the CCRA senior politicians in the UK government didn't really see climate change as such a big issue ... the economy was by far the most important thing; [climate change adaptation] slipped down the agenda and I think it's continuing to be slipped down the agenda ...
>
> [W]ith the National Adaptation Programme, I think people assumed that it would end up laying out a programme of actions by government ... and saying why we are doing these actions. It doesn't go that far ... because basically government wants to do the absolute minimum itself. It wants to be demonstrably encouraging business and community to increase its adaptive capacity, but they want to do the minimum themselves.
>
> (UK-Policy Scientist 1)

As one Australian policy scientist noted, climate change evidence and policy assessment will often have questionable legitimacy in the context of partisan politics and short-term political cycles:

> Very rarely do you see a politician who has the intestinal fortitude to embed 100 year decision-making in a 3 year [election] cycle when the science around that 100 years decision-making has a band of confidence which is very wide in the out-years;
>
> At the end of the day … if there is any doubt around the evidence and it's politically unpalatable your chances of having a politician make a difficult decision are virtually nil because they're the ones who front up [to] the cameras and have to explain to the general mass why they're making this tough decision. They won't do it if there's any doubt around some of the science.
>
> (Aus-Policy Player 2)

Pursuing the linear-technocratic model facilitates unrealistic expectations about what climate adaptation evidence can provide

In both the UK and Australia there was general agreement that impact and risk assessments often lack salience for policy players and decision-making. There were a number of inter-related reasons given for this which, in many instances, mirror the reasons given for a lack of legitimacy or credibility described above.

Most participants were in agreement that risk-based decision-making can be a useful means of conceptualising and addressing climate adaptation problems.[10] Nonetheless, interviewees highlighted a number of caveats relating to problematic decision-making frameworks based upon the linear-rational model of evidence provision. The over-riding difficulty with risk-based approaches in terms of salience relates to how they are applied and interpreted for policymaking (Aus-Policy Scientists 4, 5, 6; UK-Policy Scientists 5, 10; UK-Policy Player 4). Policy players have often failed to understand the incompatibilities inherent between traditional risk assessment methods and the character of climate change risks:

> What I saw … with the UK risk assessment was the, trying to do the perfect job of a risk assessment, doing it as a linear thing where you only do it once and so you know you've got to get the risk assessment perfect before you can think of anything else, which is crazy because I think it distracts you from thinking about solutions … or about response options.
>
> (Aus-Policy Scientist 4)

Many interviewees alluded to the difficulties of traditional risk assessment approaches for climate adaptation, which were frustrated by complexity, uncertainty and a marked divergence of values and priorities between those assessing risks and those (particularly at different levels of government) using the outputs of these assessments (UK–Policy Scientists 1, 3, 6, 7, 8, 9, 10; Aus–Policy Scientists 4, 5).

Under a linear-technocratic interpretation of the evidence-based mandate, a separation has often been mandated between technical risk assessment, the development of normative priorities based on that assessment of risk, followed by the development of policy responses.

> When the [UK]CCRA started in 2009 there was a presumption that you started with your climate projections, you go through and then it tells you what your priorities are and then you devise your adaptation planning off the back of that.
>
> (UK–Policy Scientist 1)

Yet, public and political expectations for such a clear demarcation between expert and political authority through adherence to the linear-technocratic model (Keller, 2009) are fundamentally problematic. The wicked characteristics of adaptation problems preclude adequate or objective technical assessment that would allow for a linear approach to evidence development and use for policy (Jaeger et al., 1998) (see Chapter 4).

For those with linear-technocratic expectations, these assessments lacked salience because they failed to provide the right types of information (i.e., discrete, deterministic and 'correct' answers). Conversely, those who had already tried and failed to follow the oft-prescribed linear-technocratic approach had, as a result, more realistic expectations of climate change evidence. Yet these more pragmatic policy players nonetheless considered attempts at rational assessment to lack salience. They believed that what is required in the face of wicked problem characteristics is a 'decision-based approach' that reformulates the problem (and therefore a risk assessment) in the context of explicit norms, objectives and priorities that allow policy options to be developed concurrent to an assessment of the available evidence (UK–Policy Scientists 5, 10; UK–Policy Players 1, 4; Aus–Policy Scientists 3, 4, 5):

> You really need to flip the way you're looking at the world so climate science … has been operating … before and after AR4 [The IPCC's 4th Assessment Report] … around the information-deficit view of the world which is that if I just get all this information and keep pushing it into models, eventually I'll get greater clarity and be able to resolve some of these uncertainty issues … which … just requires more money to be thrown at something which is just about impossible to resolve in a form that would be useful to decision-makers … what we really need to look at it is from a decision-centric view of the world.
>
> (Aus–Policy Scientist 3)

Many believed that risk and impact assessments need to be approached as an aide to decision-making in a way that does not expect the correct answer based on the available science. Necessarily uncertain science should inform explicit objectives and priorities and should address pre-existing issues relating to climate vulnerability and resilience. As a result of the continued pursuit of a linear-technocratic approach, however, some believed that assessments such as the UKCCRA 2012 were ultimately abstract or academic exercises on the basis of very uncertain projections of the future (UK-Policy Scientist 5; UK-Policy Players 2, 3), which only marginally informed subsequent policy decisions (UK-Policy Scientists 1, 5, 10; UK-Policy Players 2, 7; Aus-Policy Scientist 4).

In Australia, likewise, an interview participant working in Queensland state government believed that the suite of adaptation plans developed under the Bligh (Labor) government had not appropriately considered potential climate change impacts at sufficient levels of detail or across a broad-enough range of circumstances in order to be salient. This reflected, he believed, the tendency of science-led impact assessments to be unable to account for the complexities of adaptation problems: 'This state is too diverse to have a one size fits all [approach to adaptation policy]' (Aus-Policy Player 2). In the view of some Australian policy players (1, 2) this problem is compounded by the problem that:

> The political system is pushing the decision-making on how to implement the evidence to the lowest level, a level that has the least capacity and ability to be able to understand and apply it ... the planning scheme ... doesn't give them the ability to consider climate change and what its true impact is.
>
> (Aus-Policy Player 2)

Much like climate change projections, other interviewees suggested that decision-makers are only interested in climate change policy evidence if it is meaningful in terms of the timescales in which they consider policy. Because climate projections lack salience over useful timescales, however, the principal difficulty with subsequent impact and risk assessments is that they too are failing to provide salient information at temporal scales relevant for policymaking (Aus-Policy Scientist 5; Aus-Policy Player 5).

Normative choice is a necessary component of evidence for policy

Interviews indicated that *ex ante* assessment can often lack not just credibility and legitimacy but also salience by failing to explicitly address key strategic normative issues or local and contextual perspectives. The UKCCRA 2012 had low levels of perceived salience when used by policy players for the purposes of informing local government or non-departmental government institutions' policy; it had not addressed key strategic issues relevant across all scales and levels of government and had failed to provide clear evidence-based statements about national risks, vulnerabilities, and priorities that could usefully inform all levels of government (UK-Policy Players 1, 2, 7; UK-Policy Scientist 10): 'Urban Heat Island Effect, right yeah, but what does it mean?' (UK-Policy Player 1).

The UKCCRA simply rehashed oft-repeated messages concerning high level impacts that everybody was already aware of but which were not particularly salient for decision-making. Without clear statements about national risks and prioritisation, those interviewed believed that developing policy across government levels that corresponded to this evidence becomes problematic:

> I just think ... it was too vanilla! ... it didn't say anything we didn't already know ... which wasn't particularly helpful. I think because it was at a national level, it missed the opportunity to make some clear statements about the impacts of climate change at a national scale and what the key regional issues were for each part of the country and from that, kind of pull out some major issues that they wanted to really look at, at a national scale.
>
> (UK-Policy Player 2)

Notwithstanding the absence of explicit political or normative positions in the development of policy evidence, some participants believed that the information provided by risk and impact assessment, could nonetheless be a useful source of evidence for certain policymaking purposes.

A number of interview participants had used the UKCCRA as a means of promoting climate adaptation policy and its implementation at various government levels. It had become a useful document 'to wave at people' (UK-Policy Scientists 6, 7, 8, 9, 10; UK-Policy Players 3, 7) which speaks to Weiss' (1979) ideas concerning the *tactical, political* or *enlightenment* uses of knowledge for policymaking. The perceived lack of salience of *ex ante* assessments relates, therefore, to directly informing policymaking in a linear-technocratic way, even though they may still hold a degree of tactical utility as a means of garnering policy legitimacy by dint of their very existence. The lack of perceived salience, however, highlights the difficulties of technical risk assessment for policymaking, due to public and political expectations for the contrasting priorities of, on the one hand, the transparency of normative input, and on the other, the maintenance of credible expert authority in evidence-based policy.

Conclusion

The perceptions of policy players presented in this chapter confirms the theoretical analysis of prior chapters: pre-existing norms and values influence the usefulness and usability of climate science and policy evidence. Although *climate science* was generally perceived as credible by those interviewed, it lacked both legitimacy and salience because of the complex, contingent, and uncertain nature of adaptation problems and the difficulties of appropriately accounting for norms and values in the development and presentation of evidence.

Climate science is not salient because it fails to present information at spatial and temporal scales and levels of governance appropriate for decision-makers;

because it contains uncertainties that make its use difficult within a linear decision-making model; or, because it is provided in probabilistic formats that can be difficult to communicate and to understand. Climate science lacks legitimacy too because its conclusions often conflict with the norms, values and priorities of policy players or the normative expectations placed upon them (e.g., in relation to neo-liberal economic-rationalist ideals) within the evidence-based mandate.

The resulting *policy evidence*, derived (at least in part) from that climate science, can be deficient across all three criteria of credibility, legitimacy and salience, particularly for those policy players not involved in its development, or who sit at contrasting levels of government from that at which the evidence was commissioned or developed. More troubling, however, is that although policy evidence lacks *explicit* normative prioritisation as a result of linear-technocratic expectations for what good evidence is and should do, it may nonetheless be subject to covert political influence during its development and presentation.

Policy evidence is not credible because it fails to adequately account for the uncertain, complex, contingent and pluralistic nature of adaptation policy problems. These problems are very difficult to adequately and impartially apprehend from any individual governance level or scale because, apart from the considerable technical complexities and uncertainties associated with understanding social-ecological and climatic systems, vulnerability to climate is a uniquely subjective condition. Under a linear-technocratic policy model, this evidence may also be miscommunicated and/or misused as representing some objective assessment of reality. The UKCCRA was also technically deficient because it had failed to account for adaptive capacity when considering the risks from and vulnerabilities of the UK to climate change.

Policy evidence lacks legitimacy too because, despite coproduction processes designed to enhance this legitimacy, at its core it is a contingent interpretation of impact and risk that can never account for the perspectives of all policy players. As a result, it is often perceived by policy players as the politicised presentation of science. Finally, policy evidence lacks salience because of this inability to account for multiple perspectives relevant to adaptation problems, which means that it cannot inform decision-makers in the ways promoted by linear-technocratic assessment methods, particularly for those policy players at different levels and scales of governance.

It seems that policy evidence may exhibit greater salience and credibility across all levels of governance when it explicitly incorporates government values and priorities into the framing and content of its analysis. Yet climate risk and impact assessments usually fail to do so since, under linear-technocratic policy models, evidence must be seen as the impartial analysis of objective science, however redundant or incredible this expectation is. Ultimately, policy evidence lacks credibility and salience for failing to explicitly address these values and priorities of government, and lacks legitimacy because of its tendency toward politicisation. And yet, somewhat paradoxically, without some level of pre-emptive legitimacy for climate change science, it seems that these policy assessments would not be undertaken in the first place.

Nonetheless, even if governments were to be wholly explicit about their norms and priorities in the derivation of policy evidence, they could not adequately account for the values, interests and specific circumstances of all relevant policy players from the perspective of a single governance level or scale. Such is the nature of climate change, the pluralistic nature of adaptation problems and the task of understanding climate risks and policy responses that the development of evidence necessitates subjective assumptions and priorities about how we view the world, what we value and where our priorities lie. It would seem therefore that, in order to be convincing and technically credible and salient, evidence for policy must be unambiguous about its contingent nature and its subjective framing and prioritisation. In order to be considered legitimate, however, it must also attempt to be congruent with the norms and values of the broadest possible range of relevant policy players.

The difficulties of balancing these priorities cut to the core of the problematic nature of the science-policy interface and the tensions between expert and political authority therein. Although transparency of norms and politics in policy evidence may be a desirable ideal for robust evidence-based policy, doing so in practice draws unwanted attention to the basis upon which expertise assumes authority to influence policy decision-making in the first place. The presence of expert authority is desirable for both experts (to ensure they have a seat at the table) and for policymakers to rationalise their political decisions. As such, there may also be advantages to keeping norms and politics hidden within the available evidence. As I venture to suggest in Chapter 9, pursuit of linear-technocratic evidence-based policymaking may facilitate a pragmatic process that allows a range of minor yet significant political norms and priorities to pass undetected (or at least unchallenged) within the provision of expert evidence. Even so, as the cases of Queensland and the UK show, the advantages of adhering to a pragmatic process of politicisation are stronger for some cases than others, depending on how tightly a government holds to the rhetoric of evidence-based policymaking.

Either way, policy players in both the UK and Australia suggested that political preferences ultimately trump conflicting climate change evidence. This, despite an avowed allegiance to an evidence-based mandate or the use of *ex ante* policy assessment supposedly undertaken to inform policy in a robust and objective way. As one interviewee put it:

> I've never believed that evidence per se decides policy … I believe that all decisions when you get down to it are political decisions driven by particularly economic and social aspects.
>
> (UK-Policy Scientist 9)

Notes

1 In the preparation of this book, I conducted a total of 39 semi-structured interviews with climate adaptation and disaster risk management policy players in the UK and Australia.
2 This section is adapted from Tangney and Howes (2016).
3 That is, under a linear-technocratic policymaking schema.
4 Notwithstanding perceptions of policy players and the public in this regard, research suggests that the recent trend of cold winters may actually be due to anthropogenic climate change (Mori et al., 2014).
5 Due to conflicts of temporal and geographic scale between the evidence needs of policy players and climate change science.
6 After the eighteenth century mathematician Thomas Bayes. Bayesian probability is different from other types of probability in that it has a subjective component based on a prior assumed probability of occurrence, which is then updated as new observations are made (Bertsch McGrayne, 2011). In the case of climate change, prior probabilities are provided by the degree of congruence of a particular climatic outcome to the average outputs of an ensemble of GCM runs (Frigg et al., 2013b). These probabilities are not objectively assessed in terms of their propensity based on past occurrence of climatic states, or their empirical derivation based on knowledge of all possible outcomes (Crawford, 2005), neither of which is possible given, respectively, the relative non-stationarity of climate due to anthropogenic forcing and the complexity of the systems in question. Rather, Bayesian probabilities are based on the projections of a series of models of possible future climates that incorporate anthropogenic forcing using a range of subjective assumptions and assertions concerning the dynamics of the Earth's climate system (Frigg et al., 2013b; Edwards, 2010).
7 The following section is adapted from Tangney (2016).
8 Interestingly, an assessment of adaptive capacity was undertaken by the team contracted to prepare the UKCCRA 2012; however this assessment was omitted from the final outputs because of a protracted debate between DEFRA and the UK's Committee on Climate Change about what the scope and content of the UKCCRA should be (UK-Policy Scientist 1). Testimonies suggesting such significant political influence in the development of the UKCCRA also speak to ideas of competing advocacy coalitions interacting in the policy process (Sabatier, 1988) (see Chapter 2).
9 This example demonstrates not only a perceived lack of credibility of the UKCCRA due to its failure to adequately account for uncertainties and its limited use of available evidence sets, but also arguably a lack of legitimacy since a fair and unbiased knowledge production process would have accounted for a range of expert and non-expert perspectives. and interpretations of climate change risk. This example highlights the ambiguous character of Cash et al.'s (2003) knowledge systems criteria.
10 These views also align with the IPCC's fifth assessment report which has used a risk-based framework to understand and evaluate climate change risks (IPCC, 2014).

Bibliography

Adger, WN, Brown, K, Nelson, DR, Berkes, F, Eakin, H, Folke, C, Galvin, K, Gunderson, L, Goulden, M, O' Brien, K, Ruitenbeek, J and Tompkins, EL 2011, 'Resilience implications of policy responses to climate change', *WIRES Climate Change*, vol. 2, no. 5, pp. 757–766.

Bertsch McGrayne, S 2011, *The Theory that Would Not Die: How Bayes' Rule Cracked the Enigma Code, Hunted down Russian Submarines, & Emerged Triumphant from Two Centuries of Controversy*, Yale University Press, London.

Carrington, D 2014, 'Climate change will make UK weather too wet and too dry, says Met Office', *The Guardian*, 26 March, viewed 12 September 2014, www.theguardian.com/environment/2014/mar/25/climate-change-uk-weather-wet-dry-met-office/.

Cash, DW, Clark, WC, Alcock, F, Dickson, NM, Eckley, N, Guston, DH, Jager, J and Mitchell, RB, 2003, 'Knowledge systems for sustainable development', *Proceedings of the National Academy of Sciences*, vol. 100, no. 14, pp. 8086–8091.

Cash, DW, Adger, WN, Berkes, F, Garden, P, Lebel, L, Olsson, P, Pritchard, L and Young, O 2006, 'Scale and cross-scale dynamics: Governance and information in a multi-level world', *Ecology and Society*, vol. 11, no. 2, viewed 16 March 2015, www.ecologyandsociety.org/vol10/iss2/art9/.

Church, J, Hobday, A, Lenton, A and Rintoul, S 2016, 'Six burning questions for climate science to answer post-Paris', *The Conservation*, 29 February 2016, viewed 19 August 2016, https://theconversation.com/six-burning-questions-for-climate-science-to-answer-post-paris-55390/.

Commonwealth Scientific and Industrial Research Organisation [CSIRO] 2016, *Climate Change in Australia*, CSIRO & the Bureau of Meteorology, Melbourne, Australia, website viewed 20 December 2016, https://www.climatechangeinaustralia.gov.au/en/.

Crawford, M 2005, 'Three Types of Probability', in *Art of Problem Solving Introductory Topics*, presented at NASA Educators Workshop 13 September, AoPS Incorporated. Alpine, CA.

Curry, J 2013, 'Consensus distorts the climate picture', *The Australian*, 21 September, Library.PressDisplay.com/.

Dedekorkut, A, Mustelin, J, Howes, M and Byrne, J 2010, 'Tempering growth: Planning for the challenges of climate change and growth management in SEQ', *Australian Planner*, vol. 47, no. 3, pp. 203–215.

Department of Environment, Food and Rural Affairs [DEFRA] 2009, *Adapting to Climate Change: Helping Key Sectors to Adapt to Climate Change*, DEFRA, London, viewed 11 March 2013, http://archive.defra.gov.uk/environment/climate/documents/interim2/report-guidance.pdf.

Department of Environment, Food and Rural Affairs [DEFRA] 2012a, *UK Climate Change Risk Assessment: Government Report*, HMSO, Norwich, UK, viewed 6 May 2013, www.defra.gov.uk/publications/files/pb13698-climate-risk-assessment.pdf.

Department of Environment, Food and Rural Affairs [DEFRA] 2012b, *UK Climate Change Risk Assessment: Evidence Report*, DEFRA, viewed 6 May 2013, http://randd.defra.gov.uk/Default.aspx?Menu=Menu&Module=More&Location=None&Completed=0&ProjectID=15747#RelatedDocuments/.

Department of Environment, Food and Rural Affairs [DEFRA] 2013a, *The National Adaptation Programme: Making the Country Resilient to Changing Climate*, HMSO, London, UK, viewed 15 January 2017, https://www.gov.uk/government/publications/adapting-to-climate-change-national-adaptation-programme/.

Dessai, S and Hulme, M 2004, 'Does climate adaptation policy need probabilities?', *Climate Policy*, vol. 4, no. 2, pp. 107–128.

Edwards, PN 2010, *A Vast Machine: Computer Models, Climate Data and the Politics of Global Warming*, The MIT Press, London.

Environment Agency of England & Wales [EA] 2010a, *Managing the Environment in a Changing Climate*. EA, Bristol, UK, viewed 17 March 2015, www.environment-agency.gov.uk/research/library/publications/130528.aspx/.

Forino, G, von Meding, J and Brewer, G 2016, 'Climate Change Adaptation and Disaster Risk Reduction Integration in Australia: Challenges and Opportunities', in Madu, CN and Kuei, C (eds), *Handbook of Disaster Risk Reduction and Management*, World Scientific Press and Imperial College Press, London.

Frigg, R, Smith, LA and Stainforth, DA 2013b, 'The myopia of imperfect climate models: The case of UKCP09', *Philosophy of Science*, vol. 80, no. 5, pp. 886–897.

Frigg, R, Smith, LA and Stainforth, DA 2015, 'An assessment of the foundational assumptions in high-resolution climate projections: The case of the UKCP09', *Synthese*, vol. 192, no. 12, pp. 3979–4008.

Heazle, M, Tangney, P, Burton, P, Howes, M, Grant-Smith, D, Reis, K and Bosomworth, K 2013, 'Mainstreaming climate change adaptation: An incremental approach to disaster risk management in Australia', *Environmental Science & Policy*, vol. 33, pp. 162–170.

Intergovernmental Panel on Climate Change [IPCC] 2014, 'Summary for policymakers', in Field, CB, Barros, VR, Dokken, DJ, Mach, KJ, Mastrandea, MD, Bilir, TE, Chatterjee, M, Ebi, KL, Estrada, YO, Genova, RC, Girma, B, Kissel, ES, Levy, AN, MacCracken, S, Mastrandea, PR and White, LL (eds), *Climate Change 2014: Impacts, Adaptation and Vulnerability. Part A: Global and Sectoral Aspects. Contribution of Working Group II to the Fifth Assessment Report of the Intergovernmental Panel on Climate Change*. Cambridge University Press, Cambridge and New York.

Jaeger, CC, Renn, O, Rosa, EA and Webler, T 1998, 'Decision Analysis and Rational Action', in Rayner, S and Malone, EL (eds), *Human Choice and Climate Change*, Volume 3: *Tools for Policy Analysis*, Batelle Press, Columbus, OH.

Keller, AC 2009, *Science in Environmental Policy: The Politics of Objective Advice*, The MIT Press, London.

Met Office, The 2009, *Helping You Meet Climate Change Head On*, The Met Office Website, viewed 12 September 2014, www.metoffice.gov.uk/news/releases/archive/2009/warmer-future/.

Mori, M, Watanabe, M, Shiogama, H, Inoue, J and Kimoto, M 2014, 'Robust Arctic Sea-ice influence on the frequent Eurasian cold winters in past decades', *Nature Geoscience*, vol. 7, no. 12, pp. 869–873.

Murphy, JM, Sexton, DMH, Jenkins, GJ, Booth, BBB, Brown, CC, Clark, RT, Collins, M, Harris, GR, Kendon, EJ, Betts, RA, Brown, SJ, Humphrey, KA, McCarthy, MP, McDonald, RE, Stephens, A, Wallace, C, Warren, R, Wilby, R and Wood, RA 2009, *UK Climate Projections Science Report: Climate Change Projections*, Met Office Hadley Centre, Exeter, UK.

Owens, S, Rayner, T and Bina, O 2004, 'New agendas for appraisal: Reflections on theory, practice, and research', *Environment and Planning A*, vol. 36, no. 11, pp. 1943–1959.

Pitman, A 2016, 'CSIRO boss' failed logic over climate science could waste billions in taxes', *The Conversation*, 5 February 2016, viewed online 26 December 2016, https://theconversation.com/csiro-bosss-failed-logic-over-climate-science-could-waste-billions-in-taxes-54249/.

Rickards, L, Wiseman, J and Edwards, T 2014, 'The problem of fit: Scenario planning and climate change adaptation in the public sector', *Environment and Planning C: Government and Policy*, vol. 32, pp. 641–662.

Sabatier, PA 1988, 'An advocacy coalition framework of policy change and the role of policy-oriented learning therein', *Policy Sciences*, vol. 21, nos. 2–3, pp. 129–168.

Steynor, A, Gawith, M and Street, R 2012, *Engaging Users in the Development and Delivery of Climate Projections: The UKCIP Experience of UKCP09*, UK Climate Impacts Programme (UKCIP), Oxford.

Tang, S and Dessai, S 2012, 'Usable science? The UK climate projections 2009 and decision support for adaptation planning', *Weather, Climate and Society*, vol. 4, pp. no. 4, 300–313.

Tangney, P 2016, 'The UK's 2012 climate change risk assessment: How the rational assessment of science develops policy-based evidence', *Science and Public Policy*, published online 6 September 2016, https://academic.oup.com/spp/article/doi/10.1093/scipol/scw055/2525558/The-UK-s-2012-Climate-Change-Risk-Assessment-How/.

Tangney, P and Howes, M 2016, 'The politics of evidence-based policy: A comparative analysis of climate adaptation in Australia and the UK', *Environment and Planning C: Government and Policy*, vol. 34, no. 6, pp. 1115–1134.

Weichselgartner, J and Kasperson, R 2010, 'Barriers in the science-policy-practice interface: Toward a knowledge-action-system in global environmental change research', *Global Environmental Change* vol. 20, pp. no. 2, pp. 266–277.

Weiss, CH 1979, 'The many meanings of research utilization', *Public Administration Review*, vol. 39, no. 5, pp. 426–431.

7 The politicisation and scientisation of climate risk management

> I suppose it is tempting, if the only tool you have is a hammer, to treat everything as if it were a nail.
>
> Abraham Maslow

This chapter examines more closely the framing and assessment of adaptation problems through the risk-based decision-making methods that are commonly used during the development of policy evidence. In particular, this chapter explores the propensity for processes of politicisation and scientisation that were alluded to in previous chapters. As discussed in Chapter 6, policy players' views of the adaptation policy process suggest that the importance of congruent norms and values gives primacy to legitimacy, over credibility and salience, as the principal determinant of useful, usable evidence for adaptation. Yet legitimacy may be the most difficult attribute to attain in such a polarised policy arena as for climate change, due to the political and economic contingencies associated with adaptation problems and the technical limitations of the evidence available to understand them. Although the contingent nature of adaptation problems and evidence does not necessarily result in the deliberate politicisation of adaptation evidence, under a linear-technocratic model such deliberate covert expression of values and priorities is certainly possible and strongly indicated by the cases examined here.

To begin, I discuss here the history of the phenomenon of evidence-based policymaking and the theoretical development of concepts of *scientisation* by scholars such as Jurgen Habermas as a result of the resurgence of expert authority for policymaking in the twentieth century. Using case-study examples, I then demonstrate how risk-based decision-making prescriptions in particular have helped government and its agencies to justify their policy decisions. I provide specific examples of how, by attempting to shoe-horn adaptation problems into linear-technocratic decision-making models, the resulting evidence is prone to deliberate politicisation (a politicisation-by-agency). I conclude that while scientised policy debate may occur in the ways originally conceptualised by Habermas (1971), Weingart (1999) and Sarewitz (2004), the scientisation of policymaking has also been facilitated in recent years through *ex ante* risk assessment methods and the types of expert knowledge available to inform them.

As discussed in Chapter 2, constructivist scholars have long argued and demonstrated empirically that it is in the nature of science and evidence, as the outputs of social processes, to be influenced and infused by scientists' and other knowledge system participants' values and even politics (Douglas, 2009; Hoppe, 2005; Jasanoff, 1990). Given that a degree of social construction in the production of evidence appears inevitable, the resulting evidence should not be dismissed or invalidated simply because it fails to meet the unrealistic expectations of positivist ideals. Nor indeed should it be dismissed as arbitrary or unremarkable because of its constructed character. What is important for the purposes of ensuring evidence credibility (as well as legitimacy and salience), is that there is a measure of consistency between how this knowledge is derived, presented and used. In other words, it is important that there is transparency in the normative assertions made during the derivation of this knowledge and the epistemic claims made about this evidence when it is used for policymaking. Achieving such consistency, however, is no mean feat.

As I demonstrate in this chapter, the characteristics of climate-related risk assessments suggest that policy players have both the opportunity and the incentive to deliberately and covertly infuse evidence with normative and political views as a means to legitimise their political positions, or to expedite particular policy initiatives. Not only does this politicisation underline the constructivist character of evidence for policy, it raises important questions about the suitability of risk-based methods for evidence development in practice. Moreover, these difficulties challenge the extent to which it is possible to reconcile evidence credibility (so often concerned with ideals of objectivity and positivism) with its salience and legitimacy for decision-making, both of which necessitate observance of a plurality of norms and values.

Evidence-based policy: An evolving field of decision-making

Kitcher (2011) suggests that, although the idea that good decisions should be based on expert knowledge has origins dating back (at least) to Plato's 'Republic' (ca. 380 BC), it has only been since the eighteenth century or so that the natural sciences have been considered by many as the epitome of human knowledge and therefore an important constituent in public policymaking and implementation (Weingart, 1999). Despite society's dependence on science and technology for its understanding, design and provision of the trappings of modernity, however, there has often been a tension about the role of such knowledge in political decision-making. In negotiating this tension, governments have often heeded the advice of experts (for instance on public health issues such as smoking, for which there is considerable determinacy in the expected morbidity), though rarely if ever have they relinquished decision-making control. The reason for government accession in this regard is presumably because of both the instrumental and legitimating power that evidence can have for public policymaking (Weingart, 1999). However, evidence-based policymaking has simultaneously highlighted the aforementioned tensions between experts and politics and which, with the advent of applied social science, appear to have become even greater, perhaps because of the ability of the latter to address normative or political questions in a more explicit or direct way than science and engineering ever could (Marston and Watts, 2003).

Sanderson (2002) suggests that the take-off point for social science research for policy in the UK occurred in the mid 1960s, but that this relationship was problematic due to a political culture resistant to the influence of 'rational knowledge' from experts for political decision-making.[1] More recently, however, following more than two decades of 'conviction politics' under successive conservative governments, Nutley et al. (2002) and Solesbury (2001) describe how evidence-based policymaking in the UK witnessed a resurgence of interest under the 'New Labour' government of the late 1990s, who had adopted the mantra 'what matters is what works'. Davies et al. (1999) and Solesbury (2001) argue that evidence-based policy was promoted by Labour as part of a reformist, anti-ideological stance[2] in reaction to previous governments' neo-liberal convictions.[3] This new evidence-based mandate sought to demonstrate the efficacy of evidence for the resolution of policy problems, as well as policymakers' rational use of it.

In Australia, as described in Chapter 3, there has often been a general suspicion of expertise and its role for dictating environmental management practices, particularly in the state of Queensland. Kay (2011) and Banks (2009) suggest that recent ambitions toward evidence-based policy, at federal level at least, aligned with the reinstatement of the Labor Party in 2007, as newly elected Prime Minister Kevin Rudd observed that 'evidence-based policymaking is at the heart of being a reformist government' (Banks, 2009: p. 3). Yet, the clarion call from the UK concerning 'what works' was most certainly heard in Australian politics well before the arrival of the Rudd government (Holmes and Clark, 2008; Marston and Watts, 2003; Hess and Adams, 2002). Further, Banks (2009) argues that the governance ideal of evidence-based policymaking has existed in Australia since the 1980s at least, albeit principally for economic rather than social and environmental policymaking. Much as Nutley et al. (2002) argue for the UK case, however, Banks (2009) suggests that Australian federal government has been more evidence-*influenced* than evidence-based and that evidence rarely, if ever, plays a deterministic or technocratic role in political decisions.

In Chapters 2 and 4, I explored the theoretical precepts behind ideas of trans-science, post-normal science and co-produced policy knowledge and, on that basis, the types of knowledge that constitute scientific evidence for adaptation policy. The evolution of evidence-based policy, however, seems to have developed largely independently from the concurrent development of constructivist theories of science and expert authority. Only in recent times have policymakers begun to take on board the advice of the academy regarding the contingent nature of evidence for public policy problems, and an associated need for the democratisation of expertise (Maasen and Weingart, 2005). Despite increasing acceptance by governments of the need for broader concepts of what evidence and expertise are (Solesbury, 2001), rationalist models of evidence-based policymaking still prevail in ways that seem to largely conflict with constructivist arguments concerning the contingency of policy knowledge (Sanderson, 2002).

A number of key debates have arisen in the academic literature concerning the value of experts and evidence for policymaking, including those discussed in previous chapters. It is useful to consider these in relation to the two opposing conceptual models of policymaking which I discussed at length in Chapter 2. The linear-technocratic policy model, on the one hand, advances the idea that policy decisions can be wholly rationalised, and even determined directly, by available research. On the other hand, the political model, argues that policymaking is, at best, evidence-influenced but largely derived through political machinations and deliberation. These contrasting perspectives raise contentious debates about what we should consider as 'rational knowledge'. Nutley et al. (2002) suggest that although the UK government has in recent years provided a broad canvas for what constitutes evidence for policy, the prevalence of the rational model means that in practice, there is a clear hierarchy between different types of information. Quantitative scientific and social-scientific research sits at the top of this hierarchy and qualitative knowledge derived from, for instance, public consultations is situated further down:

> If knowledge operates hierarchically, we begin to see that far from being a neutral concept, evidence-based policy is a powerful metaphor in shaping what forms of knowledge are considered closest to the 'truth' in decision-making processes and policy argument.
>
> (Marston and Watts, 2003: p. 145)

If, as argued by Head (2008b), Jasanoff (1990, 2003a, 2005), and Nutley et al. (2002) amongst others, there is validity to a political model of policymaking then it seems plausible that adherence to linear-technocratic prescriptions for evidence-based policy that favour scientific expertise and quantitative knowledge over qualitative or less rigorously derived local or contextual knowledge may actually facilitate the politicisation of evidence. This is because, as discussed in Chapter 4 and demonstrated in the case-study examples in this chapter, these preferred types of evidence still often necessitate important subjective and normative decisions in their derivation that can be manipulated in the course of policy evidence development to align with political preferences.

Ex ante policy assessment tools that prescribe linear-technocratic approaches to risk assessment and that are used to fulfil governments' commitments to evidence-based policymaking encounter significant difficulties in adequately accounting for the complexity, uncertainty and divergence of opinions and values relating to climate change risks, costs and benefits and are used in ways that, I argue, are often ill-suited to the task of making robust evidence-based decisions. Moreover, they are problematic because, rather than simply serving as a heuristic device, or as a means of inoculating policymaking from the contagion of politics (Nieman and Stambough, 1998) they can actually serve the opposite purpose. In formalising notions of linear-rationality within policy appraisal as a means to standardise robust decision-making, they facilitate the politicisation of expert evidence by providing a façade of expertise behind which political choices may be disguised.

A political policymaking model, by contrast, suggests that policymakers use evidence to legitimise more than to inform their decision-making. This model may indicate, therefore, the possibility of a scientisation of policymaking whereby political choice is disguised, or debate is suppressed, through policy players' recourse to expert authority and the supposedly impartial evidence they provide. As I argue below, however, processes of politicisation and scientisation are closely related and suggest that, irrespective of which policy model may be more accurate, the evidence-based mandate can be used to manipulate the decision-influencing authority granted to experts for political ends.

The scientisation of policymaking

Perhaps one of the first scholars to discuss the scientisation of politics was Jurgen Habermas (1971) who described the distinction between *decisionistic* and *technocratic* relationships between expert and political authority.[4] According to Habermas, decisionism involves prescribed roles for politicians and expert advisers involving a clear separation between value-based (what he calls 'irrational') decision-making necessary for policymaking and as undertaken by politicians, and the legitimation and rationalisation of those decisions and their instrumentation through the use of expert evidence. In this way decisionism has strong links to linear rationalism.[5] Majone (1989) suggests that decisionism is also based on the premise that policy problems are discrete bounded issues that allow for political decisions to be justified on the basis of rationalised argument. Decisionism has its origins in the writings of Max Weber (see Chapter 2) amongst others, and suggests that government's decisions inevitably have both a technical, rational component relating to the use of evidence in the design and implementation of policy, and a purely 'irrational' component relating to the value positions of the political elite:

> As much as the objective knowledge of the expert may determine the techniques of rational administration ... practical decision in concrete situations cannot be sufficiently legitimated through reason. Rationality in the choice of means accompanies avowed irrationality in orientation to values, goals and needs.
>
> (Habermas, 1971: p. 63)

Under this model, Habermas envisages the scientisation of politics as a process by which expert authority is used in an instrumental or legitimising way to aid the irrational decision-making of the political executive. However, decisionism is flawed, he argues, since those who are supposed to impartially implement the wishes of the executive have their own values and priorities and the process of policy implementation (i.e. the choice of means) also requires value-based decisions which defy resolution through pure rationalisation.

Habermas' (1971) also pays considerable attention to an opposing model, that of *technocracy*. Like decisionism, technocracy assumes that science and policymaking are functionally separated activities. However, since policy players have increasingly sought to rationalise decision-making through the application of decision theory and systematic forms of analysis (e.g., risk assessment), Habermas suggests that the decisionist model has often competed for space with technocratic ideals. Under a technocratic model, therefore, experts would be trusted to rationalise decision-making in such a way as to make the political elite increasingly redundant:

> The politician would then be at best something like a stopgap in a still imperfect rationalization of power, in which the initiative has in any case passed to scientific analysis and technical planning.
>
> (Habermas, 1971: p. 64)

Habermas rejects technocracy on the basis that it is impossible to fully rationalise political decisions. There will always be an 'irrational' value-based component to decision-making, no matter how far technical expertise can erode the scope of political deliberation. Under technocracy then, the scientisation of politics becomes a process by which experts gain increasing control over value-preferences and political decision-making, under the guise of technical impartial control. Habermas (1971) argues for a pragmatic compromise between decisionism and technocracy in order to avoid the worst excesses of both, whereby there would be an ongoing dialogue between expert and political authority, guided by an hermeneutic understanding (i.e. interpretation) of the public's 'value beliefs' as articulated through the discourse of relevant communities (Habermas, 1971: p. 68, 69).

More recently, Weingart (1999) and Sarewitz (2004) have added considerable colour to the discussion of the scientisation of politics, particularly in relation to the resolution of environmental problems such as climate change. Weingart (1999), noting Habermas' dichotomy between decisionist and technocratic interpretations, tracks the development of governments' science-policy and the use of science for policy in the USA and Germany since World War II as a means of illuminating the relationship between experts and politics and (what he refers to as) the *scientification*[6] of policymaking that has occurred during this time. He argues that despite concerns amongst politicians and the media in the mid twentieth century that scientists were gaining increasing status and illegitimate influence over policymaking in the US (i.e., technocratic control), a democratisation of expertise subsequently occurred, resulting in ready access to expert knowledge by all participants in policymaking. What ensued was a process of scientification allowing policy players to pick and choose expert evidence, leading to 'a competition for expertise which intensifies controversies in policy-making rather than alleviating them' (Weingart. 1999: p. 152) and which delegitimises expert authority as a result. Weingart (1999) argues that the scientification of policymaking is tightly coupled with the politicisation of science; scientific knowledge is not value-free and experts engaging in policymaking are not politically neutral, particularly as the competition between experts and their evidence resulting from this scientification means that, increasingly, scientific research is used before its veracity can be adequately determined.

For the sake of simplification, Sarewitz' (2004) begins his argument by assuming that science does what it says it does: it provides impartial objective truth about the world. Similar to Weingart (1999), he argues that because there are so many disciplinary lenses through which to view the world, there are many scientific or empirically derived truths, not all of which are commensurable, and some of which outrightly conflict. Sarewitz (2004) draws attention to a crucially important characteristic of contemporary scientific knowledge in a pluralistic world, and thus the task of addressing wicked problems such as climate change (described in detail in Chapters 4 and 5); there appear to be many valid and contrasting truths and interpretations of evidence to legitimise political decisions. Sarewitz (2004) interprets scientisation as *the suppression of normative debate by recourse to the 'facts'*, which is facilitated for wicked problems by a plurality of valid scientific arguments and disciplinary lenses that can align with alternative and potentially contrasting political positions. Thus, conflicting political positions have their own sets of legitimised facts, and the choice of, or debate concerning which set to use suppresses or supplants explicit normative debate.

Sarewitz (2004) concludes that scientisation in this form occurs as a result of how science has traditionally been used in policy debates (i.e., decisionistically, to rationalise pre-existing norms and political positions). For the sake of simplicity he largely avoids discussion of political forces at play within evidence development itself, such as those described in the case-studies of this chapter. However, his argument about scientisation of environmental policy clearly implicates one form of politicisation, a politicisation-by-agency, that can occur in the choice and presentation of evidence and draws further attention to the tight coupling between concepts of scientisation and politicisation (Weingart, 1999).

Politicisation, as defined here, occurs as a result of normative or political influence in either the development or communication of science and policy evidence. I follow the arguments of Fischer (2009) concerning the legitimacy of expertise to suggest that politicisation occurs as a result of scientists extending their decision-influencing authority beyond the realms of what they can legitimately claim to have privileged knowledge about. This over-extension of expert authority can occur in two ways. First, a *politicisation-by-process* can occur during the development of trans-scientific and post-normal scientific evidence. As described in Chapter 4, this politicisation is in some respects inevitable for the *development* of climate science that must incorporate non-epistemic value judgements, and is also usually relatively benign. Second, however, a *politicisation-by-agency* can also occur, particularly during the *interpretation and communication* of science when developing post-normal scientific outputs and *ex ante* policy evidence.

Importantly, therefore, the distinction between these two forms politicisation relates to whether values influence evidence in the course of 'doing science', i.e., within the remit of scientific expertise, or whether this politicisation occurs in the subsequent interpretation of science by experts and non-experts involved in the evidence-based policy process when attempting to make science more useful and usable for decision-making. In either case politicisation occurs when values influence scientific products beyond those judgements required and circumscribed by good methodological practice (what Heather Douglas (2009) refers to as *epistemic* and *cognitive values*). The value judgements of concern here are those that either transcend or are extrinsic to agreed scientific methods, yet which are nonetheless required for the derivation of useful science or to influence experts' conclusions and communication of science in a way that promotes (intentionally or otherwise) a normative position under the guise of objective expert knowledge.

At the science-policy interface, however, and particularly during the derivation of policy evidence such as climate change risk assessment, it becomes difficult to discern where exactly evidence development ceases and political/normative influence begins. Although the distinction between politicisation and scientisation is an important one, whereby the former is something that happens to science and the latter is that which happens to policymaking, this distinction becomes difficult to discern where 'policy evidence' (as characterised in Chapters 4 and 6) is often not wholly scientific in its make-up, but rather is co-produced through some combination of technical analysis and normative prioritisation by both experts and non-experts, while still seeking legitimacy through its claims to privileged expert knowledge. Under such circumstances, these phenomena are so tightly coupled that, I argue, Weingart's (1999) and Sarewitz' (2004) characterisation of scientisation is inadequate on its own because they assume that expert evidence is produced solely by experts. Both scholars leave unanswered the question of what happens to policymaking when non-expert policy players are afforded the opportunity to co-produce policy evidence which, under a linear-technocratic schema, is construed as objective expertise. In such circumstances, I argue, another mechanism of scientisation prevails whereby non-expert policy players can politicise evidence within the functional bounds of expert authority (rather than through their subsequent interpretation and communication of that expertise during the policy process) in order to scientise.

It is at the science-policy interface for wicked policy issues like climate adaptation that non-expert policy players have the opportunity to deliberately over-extend expert authority on experts' behalf. This over-extension occurs as a result of contradictory expectations for both an adherence to the linear-technocratic model to ensure credibility on the one hand, and for the democratisation of evidence production for policymaking to ensure political acceptability (i.e., legitimacy) on the other. As the cases of the UK Climate Change Risk Assessment (UKCCRA) in the UK, and Q100 assessments for the city of Brisbane described below demonstrate, normative prioritisation can quite easily become wrapped up in technical considerations of environmental management and policy players can deliberately manipulate those considerations to ensure expert evidence tells the story best aligned with pre-existing norms, policies and politics.

Interestingly, Weingart (1999: p. 157) eschews the idea that the boundary between expert and political authority has been blurred because '[this idea] incorrectly assumes that the functional differentiation between science and politics disappears'. However, the case-studies described below indicate that during the co-production of policy evidence non-experts do have the opportunity to undertake normative/political prioritisation under the guise of impartial, independent expertise. The difference between this form of scientisation and those identified by Weingart (1999) and Sarewitz (2004) is that, in the former case, expert authority is manipulated through the actual development of evidence, whereas Weingart (1999) and Sarewitz (2004) argue that expertise is manipulated through the choice and communication/presentation of evidence. Alternatively, of course, experts may over-extend their own authority by annexing trans-scientific or policy questions as technical matters when they are encountered and can be subsumed during the development of climate science and policy evidence.

As the case studies below demonstrate, the scientisation process consists of the political executive saying: 'Our policy is informed by and coherent with the facts that experts provided', while those facts were actually derived not just from scientific research or informed expert judgement, but through knowledge co-production processes between experts and non-experts. During this co-production process, values and priorities are negotiated through some combination of problem framing, characterisation and technical modelling and assessment, through the choice of which evidence to use and which experts to listen to. I argue that during this co-production process science can be politicised by various policy players in the course of creating an internally consistent body of policy knowledge in line with government politics. This, in turn, results in the scientisation of policy because this politicised evidence has covertly answered and suppressed substantive normative debates and is used to legitimise policy positions through recourse to these 'facts'. Adaptation policy, therefore, becomes scientised in two ways (see Figure 7.1):

1   when seeking to legitimise decisions through the choice of one set of facts, or interpretations of facts, over other available and equally valid sets or interpretations (Weingart, 1999; Sarewitz, 2004); or,

2   when seeking to legitimise decisions through politicisation-by-agency, during the development of a single available set of evidence by either experts or non-experts, to ensure that government's rational recourse to the 'facts' is absolute.

Both forms of scientisation involve an attempt by partisan players to shut down specific areas of normative or political discussion through recourse to supposed objective truth. Through the co-production of policy evidence, however, the distinction between politicised evidence development and politicised evidence choice or communication (i.e., scientisation) has become so slight as to demonstrate the extent of overlap that has developed between science and politics during evidence-based policymaking. Politicisation and scientisation are not just two sides of the same coin; contemporary co-production processes for policy evidence development mean that the functional differentiation between science and politics has indeed become blurred. Figure 7.1 provides a visual aid to understand these interactions between processes of politicisation and scientisation at the science-policy interface. What is important to note in this diagram is that both experts and non-experts may promulgate values on either side of this interface, through the development of both evidence and evidence-based policy, or either under the guise of expertise or through bureaucratic politics. I argue that the interface between knowledge and values

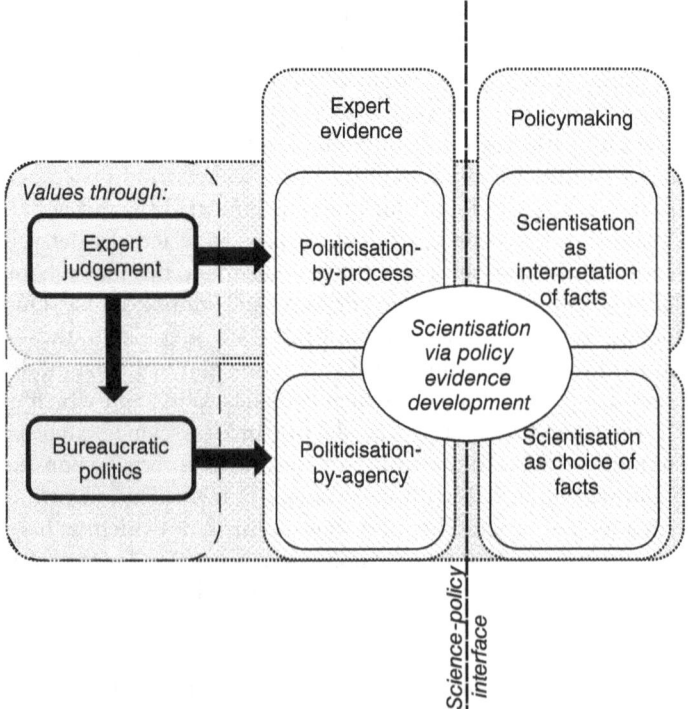

Figure 7.1 Processes of politicisation and scientisation at the science-policy interface

has become so porous as a result of the advent and development of policy evidence that both expert and political authority may be advanced from either side or used by both experts and non-experts as a means of achieving their normative priorities.

Risk-based decision-making for climate adaptation in Australia and the UK

Under the linear-technocratic model, the concept of risk or risk-based decision-making holds considerable allure. As described by Renn (2008) and Beck (1992), amongst others, contemporary society has become shaped and preoccupied by considerations of risk, as the trappings of modernity have become a potential source of its own destruction.[7] The unintended negative consequences of technological advance and collective action can outweigh its intended benefits, and therefore risk has become a sort of currency in the global economy whereby it has become a principal determinant of capitalist expansion:

> The gain in power from techno-economic 'progress' is being increasingly overshadowed by the production of risks … unlike the factory-related or occupational hazards of the nineteenth and the first half of the twentieth centuries, these can no longer be limited to certain localities or groups, but rather exhibit a tendency to globalization.
>
> (Beck, 1992: p. 13)

Lash and Wynne (1992) argue that, although risk is a multidisciplinary concept that provides the illusion of pluralistic research and debate, the dominant discourses of risk are instrumentalist and reductionist in their outlook, thus aligning with linear-technocratic perceptions of policymaking.

In this vein, Renn (2008) argues that the contemporary definitions of risk contain three elements:

1 outcomes that can impact on what humans value;
2 the possibility of occurrence;
3 a formula to combine both elements (e.g., Risk = Likelihood X Consequence).[8]

Indeed, perhaps it is because of this empirical formula that Lash and Wynne (1992) argue that, just as the types of evidence for policy are often arranged by government hierarchically (Marston and Watts, 2003), so too are the types of expertise available: 'technical experts are given pole position to define agendas and impose bounding premises *a priori* on risk discourses' (Lash and Wynne, 1992, p. 4). Yet, as discussed in relation to climate change in previous chapters, risk is also clearly a social construction that depends on personal and socio-cultural interpretations of what is important and what we value, which defies any 'objective' instrumental calculation by technical experts. Risk acceptability is fundamentally a political issue (Douglas and Wildavsky, 1983). It may be inevitable, therefore, that technical risk assessors are influenced in their assessment of risk by cultural values, by competition for resources and prevailing economic priorities, and not just by the format and availability of scientific research and other forms of evidence (Renn, 2008). In Chapter 8 I argue that, although the risk-based approach to climate change is widely endorsed (IPCC, 2014, p. 1772), due to its reductionist, positivist underpinnings and the propensity for the politicisation of evidence allowed as a result, it is not as universally appropriate a framework as it is often promoted to be. Indeed, I argue that the risk-based approach should be subsumed under a framework that allows for greater normative transparency in the development of strategic evidence for climate adaptation policy.

In order to demonstrate the contingent nature of risk assessment for climate adaptation, I describe here two examples, the first from the UK and the second from Queensland. These cases have both used an empirical risk formula in ways that have combined research science and prevailing norms and politics in the development of policy evidence to rationalise policy decisions, under a façade of impartial technical analysis. These case-studies provide useful examples of the processes of politicisation and scientisation that can ensue when using reductionist risk-based methods to understand necessarily contingent adaptation problems. These examples help to address the question of whether *ex ante* policy assessment, and risk assessment approaches in particular, may be prone to politicisation, beyond that which is inherent and inevitable in the social construction of otherwise robust climate change evidence.

The UK Climate Impacts Program – Risk framework[9]

The UKCCRA was developed using a method published by the UK Climate Impacts Programme (UKCIP) in 2003 (DEFRA, 2009; Willows and Connell, 2003). However, this method draws heavily on decision-making frameworks that have been in development by UK government departments and agencies over many years, following a format very similar to risk assessment methods used for evidence-based policymaking for the natural environment dating back at least to 1995 (DEFRA, 2011a). The UKCIP framework, therefore, follows a traditional rationalist schema that assumes that climate scientists and other experts can apprehend reality with sufficient accuracy to impartially assess climate risks and inform judgements about those risks for decision-making. The method gives explicit priority to scientific and technical evidence for informing these decisions:

> The assessment of climate risks is a complex undertaking **that must support judgements** and decisions concerning appropriate future courses of action. It requires a combination of **scientific and technical knowledge** … of society's tolerance and acceptance of risk, and the costs and benefits of different courses of action.
>
> (Willows and Connell, 2003: p. 3) [Emphasis added]

The framework assumes that climate adaptation problems are tractable, that they can be defined 'correctly' (ibid, p. 7, 10) to facilitate 'good practice in decision-making' (ibid, p. 6), allowing 'robust decisions' (ibid, p. 5) to be made using the 'best information available' (DEFRA, 2012b, p. V). Implicit throughout this guidance is the assumption that adaptation problems can be unambiguously defined, to the satisfaction of all (or most) relevant stakeholders, using science and other technical knowledge. Judgements and policy decisions are considered, not so much a part of the risk assessment process itself, but something which can be usefully informed by the assessment of risk in a linear and transparent way. The framework suggests that robust decisions are essentially rational ones: they are informed by sufficient objective expert knowledge about the nature of risk to be able to rationally weigh the costs and benefits in order to identify and implement 'good adaptation options' (Willows and Connell, 2003: p. 4).

As I argue here and in previous chapters, such assumptions about the tractability of adaptation policy problems and the objective nature of the available evidence are difficult to justify in practice. The expectations and information requirements for rational policymaking cannot be met when addressing most contemporary policy issues. Rittel and Webber's (1973) use of the term 'wicked' to describe problems that display a high degree of uncertainty, complexity and divergence of values and objectives amongst decision stakeholders appears particularly apt to describe climate adaptation issues (Head, 2008b). There are few, if any, definitive answers to climate adaptation problems, particularly when it comes to human interactions with the environment and the evidence that exists is open to subjective, normative influence, selection and interpretation (Oreskes, 2004; Sarewitz, 2004).

More troubling, however, is the idea that climate risk, which constructivist scholars argue is a fundamentally normative concept to begin with (Renn, 2008; Douglas and Wildavsky, 1983), can be impartially apprehended by government in a way that provides an adequate summation of any given policy problem to allow rationalised political judgement. As demonstrated by my analysis of climate-related evidence in Chapter 4, and the perceptions of policy players regarding adaptation evidence described in Chapter 6, this rationalist assumption is fundamentally misguided. Climate risk cannot be defined 'correctly', any more than it is possible to define objectively 'good adaptation options'. The assessment and management of climate risks necessitates important contingent judgements by those tasked with undertaking it and, I argue, may curtail subsequent explicit political debate due to the propensity for these assessments to become politicised, and for policy players to use this information in a scientised policymaking process.

As described in Chapter 6, the resulting difficulties of using the UKCIP risk assessment framework were keenly felt amongst those who participated in the development of the UKCCRA. Participants described ongoing tensions between the rationalist risk framework prescribed by UKCIP and the realities of apprehending the risks from climate change at a strategic level. These difficulties related to:

1   The sheer complexity and uncertainty of understanding risks at a national level that would provide unambiguously correct answers or that could avoid descending into vague and largely meaningless generalities.

2   The difficulties of using climate science to assess risk in a linear 'top-down' or deterministic way; that is, using climate science to understand future hazards with sufficient precision and accuracy to effectively understand climate change risks to chosen receptors.

3   The difficulties of avoiding the influence of prevailing norms and values or the influence of cognitive bias in the risk assessment process. For instance, confirmation bias was suggested by a number of interview participants involved in the UKCCRA whereby experts identified risks that aligned with their pre-existing academic priorities and expertise.

4   The pressure placed upon the assessors to deliver politically acceptable results by public servants working in a wide range of ministerial departments, who were then asked to sign-off on the assessment and its conclusions before it was presented to parliament.

Testimony suggested political pressure on the risk assessment process that not only correlates with participants' perceptions of failing credibility, legitimacy and salience of policy evidence, but also suggests the possibility of deliberate politicisation of co-production processes that claimed to adhere to rationalist principals of impartiality and privileged expertise (UK-Policy Scientists 1,2,3,10).

Participants in the UKCCRA indicated that the assessment was, to some extent, deliberately influenced by prevailing politics in order to ensure tacit acceptance of particular norms and values. As one policy scientist noted of the UKCCRA's assessment process:

> You kind of needed to be robust, otherwise you were at risk of being kicked all over the field, because it's an area that evokes quite strong views … it meant that it was a very political environment.
>
> At the stage we were defining the problem there was very little interest [from ministerial departments]; once we'd done the analysis and produced the initial report, suddenly there was interest and suddenly they wanted something different … we did quite a lot of second and third phase analysis.
>
> It started to get quite brutal from a contractor point of view … we misread it, we treated it as a technical contract, [but] the technical work was probably less than 40% of the contract, and most of it was to do with stakeholder engagement with a particular emphasis on government departments rather than the original intention which was for a much wider stakeholder engagement … and then producing a report which met departments' needs and that was completely different to producing a report that stated the technical case.
>
> (UK-Policy Scientist 2)

The policy scientist in question described how the initial technical report was deemed unacceptable by ministerial departments affected by the conclusions of this analysis:

> At the end of the main, the first reporting phase, there was a realisation by government that the reports, the technical reports were not what they wanted so there was a new phase negotiated which was all about producing the reports you now see on the website, which was, I mean they went through about four or five edits, and there were thousands of comments … And that was kind of in parallel with a change in their approach to stakeholder engagement … It was certainly our intention to go with the initial technical result and discuss it with stakeholders but that's where there was a change of direction and a real concern by the government departments that they didn't want to go public on this until it had been laid before parliament.
>
> The technical report that they initially produced essentially cranked the handle on the methodology and produced some results … and then you've got to interpret it, but here it was a case of interpreting it and ensuring you kind of addressed the interests of each department, and that was a mix of not saying things that would scupper existing policy as well as having some sight of where they're trying to get to and making sure that there was information there.
>
> (UK–Policy Scientist 2)

Another prominent UK policy scientist (10) agreed with this view of government intervention and believed that the UKCCRA had only presented the risks that were deemed politically acceptable for government to recognise. This scientist believed the assessors and their government patrons had deliberately excluded the issue of extreme sea-level rise because of the political ramifications of doing so. The policy scientist in question (a national expert in sea-level rise and associated flood risk on the urban environment) believed that a credible, legitimate risk assessment would have addressed the full range of risks from sea-level rise given the considerable scientific uncertainties regarding their likelihood, and due to the potentially catastrophic consequences on particular locations and regions of the country that could arise from step-changes in sea-level due to Greenland and Antarctic ice melt:

> I came across quite a few civil servants who said, well there's no appetite for this … so we're not going to tell it … we're not going to do anything about it … even if you say well this is the implication, it's very difficult, but we should do something about it, [and] if we don't X will happen; but if you don't even get to that stage how can you possibly blame the politician?
>
> (UK–Policy Scientist 10)

Such political interference, however, was alluded to by a number of others in relation to the derivation of the broader risk assessment. One local government policy player who had been involved in the UKCCRA suggested that it had failed to address the 'real' risks from climate change because: 'To be perfectly honest the politicians weren't interested in it' (UK-Policy Player 2). Another prominent UK policy scientist described the UKCCRA as: 'The political presentation of science' (UK-Policy Scientist 5), while another interviewee stated:

> I'm less convinced [the UKCCRA] does represent a sort of [expert] consensus or even a biased consensus on the basis of who shouted loudest, I think it was between who won the contract and who was pulling the strings in DEFRA[10] at the time.
>
> (UK-Policy Scientist 7)

The case-study of the UKCCRA is a useful example demonstrating the limitations of risk-based approaches to adaptation policy evidence. Using a method long-prescribed by liberal democratic governance, the UKCCRA initially produced information that was deemed largely unacceptable by those working in government who were under pressure to adhere to existing policy positions. The assessment was, however, highly adjustable because of the many and varied complexities and uncertainties associated with assessing climate hazards and impacts, which meant that it could not provide irrefutable or definitive answers as per the expectations of the linear-technocratic model. It seems unsurprising, therefore, that this supposedly independent, impartial technical analysis had failed the tests of credibility, legitimacy and salience, in particular for those policy players at different jurisdictional and geographic scales and levels of governance.

This case-study shows that there are a range of potential sources of normative/political influence and bias associated with the interpretation of research during *ex ante* policy assessment. The problem with risk assessment of this kind, I argue, is not that it contains subjective and normative influence. Rather, it is that it is portrayed by government as impartial and largely technical evidence to legitimise political priorities. The combination of technical analysis and political influence from the various advocacy coalitions involved in the development of the UKCCRA described above, was portrayed by the relevant government Minister to a parliamentary and public audience as a rational, impartial and comprehensive assessment of the available science to understand climate change risks (see DEFRA, 2012a,b):

> The UK is at the forefront of climate science. Whilst the future is highly uncertain, we can use the **best scientific evidence** available alongside well established risk-based decision approaches **to assess risks and decide how to respond** ... The CCRA Evidence Report is a world-class **independent research** project that analyses the key risks and opportunities

that changes to the climate bring to the UK. It **provides a baseline** that sets out how climate risks may manifest themselves ... The baseline of the CCRA Evidence Report allows Government and others to assess the extent to which our actions and plans are climate resilient.

(DEFRA, 2012a: p. 3) [Emphasis added]

It presents the **best information** available on the vulnerability of the UK to climate change, **identifies notable risks and opportunities and gaps in our current understanding** of climate risks.

(DEFRA, 2012b: p. V) [Emphasis added]

In reality, the outputs of the assessment appear to have been anything but 'independent' and its description as the 'best information available' holds more than a whiff of rationalist sentiment and therefore seems audacious given that it was adjusted by various advocacy groups within government so that it aligned with political priorities. This process of politicisation has important implications for the viability of risk-based decision-making methods and their propensity for facilitating the scientisation of policy debates and policymaking.

Floodplain management in southeast Queensland: The use of Q100 metrics to determine urban planning policy[11]

In this section I summarise the complex history of the development of risk-based methods that have been used to determine urban planning controls in Brisbane, southeast Queensland over the past 40 years. I describe how these methods have effectively concealed political decisions about communities' flood risk, vulnerability and exposure within an apparently impartial technical assessment, and I argue, indicates a process of politicisation at work in the development of climate risk management policy by Brisbane City Council. This history is complex and in some aspects incomplete, but it nonetheless demonstrates how flood risk assessment methods that dictated the tolerable level of urban development in the Brisbane River floodplain between 1976 and 2011, were based on politically motivated assumptions about the operation of the Wivenhoe Dam in order to maximise urban development (see Figure 7.2; see also Box 4.4, Chapter 4). The normative assumptions required for the operation of the dam were influenced by political motives that proved to be inadequate for the purposes of flood risk management, and in particular, for minimising the flooding events in southeast Queensland in 2010-2011 (QFCI, 2011, 2012). These assumptions were instrumental in adversely affecting the vulnerability and exposure of communities in Brisbane to extreme flooding events.

What is interesting about this case-study is that, although it relates directly to the issue of climate adaptation policymaking, it avoids grasping the thorny issue of climate change and much of the associated indeterminacy of technical risk assessments faced during the UKCCRA, described above. Yet this assessment

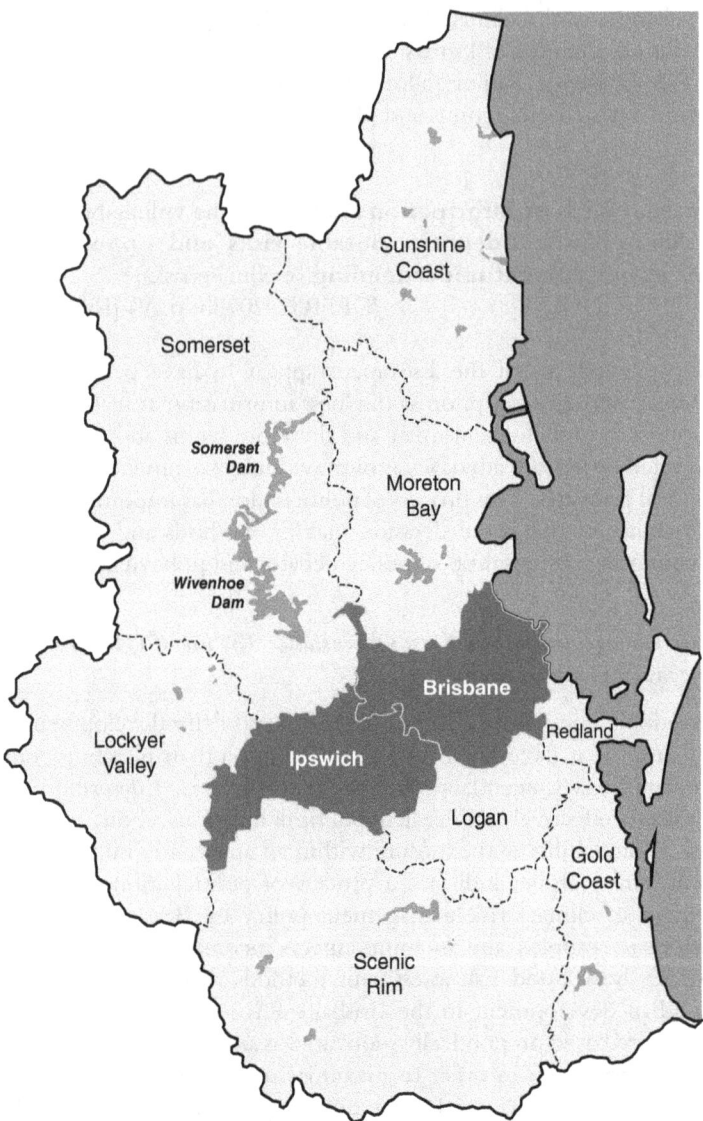

Figure 7.2 Southeast Queensland showing the local council jurisdictions of Brisbane and Ipswich alongside the Wivenhoe and Somerset lakes/dams

of risks was nonetheless hampered by difficulties of uncertainty, complexity and divergent norms and values that allowed supposedly expert evidence to become subject to political influence. This case demonstrates how, even when ignoring the spectre of climate change, understanding bio-physical and social-ecological systems still requires a balance between evidence credibility, legitimacy and salience (or between socio-cultural and technical reasoning) for the purposes of policymaking.

Climate risk management challenges for the city of Brisbane

In December 2010 and January 2011, an unusually strong La Nina event and prolonged intense monsoonal rainfall caused extreme flooding in the Brisbane River catchment, Lockyer Valley, and surrounding areas of southeast Queensland. The flooding resulted in the deaths of 35 people and an estimated $5 billion worth of damage (Queensland Police, 2011; QFCI, 2011: p. 20). The La Nina event in question and its associated rainfall were forecast by the Australian Bureau of Meteorology, which briefed the Queensland government of the potential for heavy rainfall in advance of the floods (QFCI, 2011). As described in Chapter 3, the Brisbane area has a long history of extreme climate and weather events. The 2011 floods were preceded by a prolonged 'Millennium Drought' in southeast Queensland between 2001 and 2009, while the last major flooding event in Brisbane on a par with 2011 had occurred in 1974.

In the midst of these water resource and flood risk management pressures, state and local government were also seeking to manage a period of unprecedented population growth in the southeast Queensland region over the preceding decades. This growth had exacerbated the pressures upon water supply during the Millennium Drought (Seqwater, 2015) and so provided sufficient motive for BCC to continually resist changes to its urban planning flood risk management control levels, as determined by the estimation of the 'Q100' metric for Brisbane. Q100 is the term for a commonly used Annual Exceedance Probability derived to inform flood risk management practices in Australia. It denotes a flooding event that is likely to recur once every one hundred years and is therefore often considered synonymous with a 1 per cent probability of recurring annually (BCC, 2013; QFCI, 2012: 40).

Brisbane has traditionally had one of the lowest levels of urban density of over 50 comparable US, Canadian and Australian cities (Searle, 2010). Perhaps as a result of decades of unchecked urban sprawl, each of Brisbane's City Plans between 1976 and 2014 placed a strategic priority upon inner city redevelopment, consolidation and/or increased density of particular inner city suburbs (BCC, 2014, 2000; Searle, 2010; Stimson and Taylor, 1998; Heywood, 1986; Simsion et al., 1979). This re-zoning is noteworthy because it has occurred despite many of these locations being in areas of high flood risk. Any upward adjustment of Q100 would likely expand these flood risk areas of the city (City Design, 1999). When considering BCC's priorities for managing flood risk in the period 1976–2011, therefore, it is important to understand that the designation of flood risk controls based on Q100 could impact upon the delivery of the City Plan, upon concerns of property developers and owners, as well as being problematic in terms of council members' electability.

These ongoing pressures upon BCC to manage flood risk for the communities they represent, I argue, provided significant disincentive to expand flood risk zones beyond their existing bounds and, during a period (1976–2011) which saw considerable urban redevelopment, consolidation and densification, such disincentives became increasingly acute. As discussed below, an independent investigation revealed how BCC have, at best, failed to coordinate their political priorities with Queensland State government, and at worst, may have even deliberately avoided adopting a higher, though more accurate, Q100 assessment for the purposes of urban development control.

In the aftermath of the 2010-2011 floods, the Queensland government established an independent Commission of Inquiry into state and local governments' DRM provisions. The commission's interim and final reports focused much of their analysis on the operation of the Wivenhoe and Somerset dams (QFCI, 2011, 2012). As discussed in Box 4.4, Chapter 4, the operation of these dams necessitates normative, political decisions in a zero-sum game concerning the relative priority given to the management of the opposing risks from flooding and drought. However, while political decisions about the dual and conflicting roles played by the dams directly influenced the scale of the flooding the cities of Brisbane and Ipswich were exposed to in 2010–2011 (Heazle et al., 2013), I argue that these political decisions also indirectly influenced the level of vulnerability of communities to flooding extremes during this period. For the purposes of this analysis, I summarise here the commission's reporting of the development of a flood risk management strategy for the city of Brisbane in order to illustrate the political influence that can be present in governments' design and use of technical evidence for risk-based decision-making.

Brisbane City Council's flood risk management strategy: The assessment of Q100

The Queensland Floods Commission of Inquiry's final report (QFCI, 2012) tells the official record of how flood risk management strategies used in the development of planning schemes for Brisbane and Ipswich City councils were derived. The strategies of both councils used a risk-based decision-making approach that used Annual Exceedance Probabilities (AEPs) as metrics of flood risk. According to the Commission of Inquiry, AEPs have been used in Queensland since the 1970s to provide an indication of the probability of flooding at specific locations (QFCI, 2012). The derivation of Q100 levels through hydrologic/hydraulic modelling was often highlighted during this period for a specific location to communicate urban planning and development limits (e.g., at the Port Office gauge in Brisbane) (QFCI, 2011, 2012), but Q100 levels were derived at various points across a catchment. This technical assessment allowed local government to specify through their planning schemes where, at what elevation on the floodplain, and what kind of development would be permitted in flood risk-prone areas. From 1976 to 2011, Brisbane City Council relied on a Q100 of 3.7 metres measured at the Port Office gauge in Brisbane (BCC, 2014; QFCI, 2012). During this time the council received multiple estimates from expert engineers of the Q100 level, ranging from 3.16

to 5.34 metres. Yet, the level of 3.7 metres adopted in 1976 and based on the peak height that would have been reached in the 1974 flood – if it had been mitigated by the Wivenhoe Dam[12] – was still in use by the time of the flooding events of 2010–2011 (QFCI, 2012).

The first study to establish a more accurate Q100 estimate was commissioned in 1996 and delivered to Brisbane City Council in 1998 with a figure of 5.34 metres. However, BCC's Water Resources Manager had reservations about the method used and commissioned a review of this estimate by an expert hydrologist. Amongst other technical reservations, the Water Resources Manager had questioned the assumptions made in the derivation of the estimate that the Wivenhoe and Somerset dams would be at 100 per cent of full supply level[13] at the beginning of any flooding event (QFCI, 2012: 49; City Design, 1999). The expert reviewer agreed with the council's concerns and concluded that 5.34 metres was an overestimate.

> conservative assumptions in key input variables point to the likelihood that the magnitude of the Q100 obtained in this Study ... is an over-estimate.
>
> (Mein et al., 2003: 26, 27)

BCC then commissioned a second estimate to address these concerns. Produced by in-house experts in June 1999, this second estimate gave a Q100 of 5.0 metres (City Design, 1999). BCC's Water Resources Manager was again unhappy with this figure, since its derivation had not adequately addressed the concerns of the reviewer of the original estimate. A third estimate was commissioned and produced in December 1999 with a Q100 of 4.7 metres, and again the Water Resources Manager was unhappy with this assessment (QFCI, 2012: 49).

Strangely, neither the commission's report (QFCI, 2012) nor a 2003 expert review panel report on past Brisbane flood studies (Mein et al., 2003) reveal the precise reasoning for why the latter two assessments were considered inadequate; however, both intimate that – much like the first estimate – BCC's concerns were due, in large part, to the precautionary assumptions made about the operation of the Wivenhoe and Somerset dams and, in part, due to other technically inappropriate assumptions relating to rainfall runoff variables used in the models (QFCI, 2012: 49; Mein et al., 2003). The expert reviewer of these estimates nonetheless noted:

> the amount of flood storage in these dams is very significant relative to design runoff values, so the correct simulation of these dams (and their operation during events) is of paramount importance.
>
> (Mein et al., 2003: 9)

Although the commission's reports are not explicit about BCC's precise reasoning for rejecting the second and third estimates, a report submitted by its team of in-house experts responsible for the second Q100 estimate clearly retains and justifies the original contractors' precautionary assumptions about the operation of the Wivenhoe Dam (City Design, 1999: 7). BCC's Water Resources Manager was clearly unconvinced by their rationale and made no adjustment to Brisbane's Q100 level or to its planning scheme on the back of these first three estimates as he considered that further work on the Q100 estimate and the city's flood study was needed (QFCI, 2012: 49).

Brisbane City Council decided to consult the state Department of Natural Resources and Mines, the Bureau of Meteorology and the Southeast Queensland Water Corporation[14] at a workshop held in October 2000. The purpose of the workshop was to decide on a most appropriate method for the estimation of Q100. Participants at the workshop were informed that new studies were being commissioned by the state department that would clarify the government's preferences (concerning the relative priority given to water resources versus flooding) in relation to likely releases from the Wivenhoe Dam in the event of major flooding, and which would produce a Q100 estimate closer to the original estimate of 3.7 metres.

Although the results of this study were expected in December 2000, the department's analysis was not made available to the council until June 2003, despite some intense media scrutiny in the intervening period due to the ongoing absence of an up-to-date Q100 (QFCI, 2012: 49). Because of this public scrutiny, the lord mayor of Brisbane at the time commissioned an independent review panel to investigate the results that would eventually be provided by the state department, and to oversee a final, definitive Q100 estimate for Brisbane (Mein et al., 2003). Notably, the review panel was chaired by the expert who had reviewed the original study in 1998. More intriguing still, the commission reveals how the terms of reference for the review panel included the sentence:

> Even if the Q100 changes from 6800 m3/s [i.e., 3.7 metres at the Port Office gauge], it is likely that the Development Control Level will remain the same as is currently used in the Brisbane City Plan.
>
> (QFCI, 2012: p. 50)

The senior engineer in the water resources branch of the council who had written these terms of reference told the Commission of Inquiry that he intended this statement to mean that if the Q100 level was found to be lower than the existing level, planning control levels would not be correspondingly lowered (QFCI, 2012, p. 50).

The commission appears to have accepted this testimony at face value. Yet, given their rejection of three prior estimates, the political and planning ramifications of a final Q100 estimate that was considerably higher than 3.7 metres must also have been considered and was likely to have been an unattractive proposition for the council's management of a rapidly expanding city. It seems reasonable to assume that BCC's engineer, when writing the terms of reference, would understand that any new and impartial Q100 estimate taking a risk-based approach could be higher than 3.7 metres, even when using less precautionary assumptions about the operation of the dams. It seems worthy to note, therefore, how the supposed technical inadequacies of previous models and BCC's preferred assumptions regarding the operation of Wivenhoe Dam during a flooding event, aligned to avoid the obvious political and economic ramifications that are likely to have resulted from a more precautionary assessment of Q100.[15] The commission's report describes how this fourth set of revised modelling results was provided in September 2003. The review panel and its contractors determined that the best estimate of Q100 was 3.3 metres, which was subsequently adjusted to 3.51 metres, and adjusted yet again in February 2004 to 3.16 metres.

Although the commission's reports do not specify what assumptions were used for these three estimates, the 2003 expert review panel report suggests that they were based on assumptions that the Wivenhoe and Somerset dams could reduce downstream peak flood flow rates by about 35 per cent on average[16] (Mein et al., 2003: 15). This assumption contrasted with the operation of the dams during the 2010–2011 floods, in advance of which the Wivenhoe Dam was at full supply capacity and therefore had minimal peak flow reduction capacity. Modelling conducted by Seqwater (the dam operator) immediately after the 2011 floods suggested that the flood mitigation capacity assumed by BCC for their derivation of Q100 would have required a significant pre-emptive reduction below the full supply capacity of Wivenhoe Dam (QFCI, 2011: 51).

BCC recommended to civic cabinet that the existing planning control be maintained at 3.7 metres, as per the rather suspicious terms of reference cited above (QFCI, 2012: 50). Despite the review panel's report recommending that Monte Carlo analysis be undertaken at a later date to address the uncertainties in the model, no such recommendation was made by BCC to its governing cabinet. The recommendation to maintain the existing planning control was approved by elected members of council in December 2003. BCC concluded that Monte Carlo analysis was beyond best practice, since the existing planning control level (3.7 metres) was over and above the three Q100 levels provided by the review panel between September 2003 and February 2004 (QFCI, 2012: 51).

Flood risk management: A political approach to an 'evidence-based' decision?

The story recounted above, as investigated by the Commission of Inquiry suggests that significant political forces were at play in the technical assessment of environmental risk by local government. BCC's initial assessment of Q100 (3.7 metres) in 1976 was based on an assumption of the flood mitigation that would have been provided by a Wivenhoe Dam, had it been in place during the flooding events of 1974 (QFCI, 2012: 48). Although the Commission of Inquiry did not specify what precise operating conditions for the dam were assumed and that had underpinned this estimate, BCC's final estimates of Q100 in 2003-2004 appear to have been based on dubious assumptions about the flood mitigation capacity of the dams in practice. Given the zero-sum game (between prioritising water resources versus flood risk management) involved in the dam's operation, any assumption concerning the operation of the Wivenhoe and Somerset dams constitutes a very significant political decision on the part of BCC where it is used to inform subsequent metrics of flood risk management and urban planning.

When BCC sought to update their Q100 estimate in 1996, they were unhappy with the operating assumptions taken for three consecutive estimates by both technical contractors and in-house experts concerning the dam's operation. The contractors' initial precautionary assumption, that the dam would be at full supply level and thus have the least potential to alleviate flooding, was deemed unacceptable as a baseline from which to assess the Q100 level for Brisbane's planning scheme. Yet it was this exact operating condition[17] that was in place during the 2010–2011 floods (QFCI, 2011: 47; Heazle et al., 2013).

Clearly then, questions arise about the credibility and legitimacy of the final Q100 estimates derived in 2003–2004, given:

1   The underlying assumption by BCC about the flood risk management strategy underpinning the operation of Wivenhoe Dam and used to assess Q100 was ultimately incorrect – it conflicted with the actual operation of the dam during the 2010–2011 floods.
2   The obvious potential motives of BCC in maintaining a planning development control level of 6800 m³/s (3.7 metres) at the Port Office gauge, Brisbane, whereby raising the level to account more realistically for the operation of the dams would likely have a considerable impact on property values and further development in the floodplain (City Design, 1999); as well as the circumstantial evidence revealed by the Commission of Inquiry suggesting that BCC has continually avoided its adjustment in the period 1996–2011 to a more precautionary level (QFCI, 2012: 49, 50).

It appears that the political decisions made by state government concerning the Wivenhoe Dam and by BCC concerning the city's flood risk tolerance conflicted in ways that ensured that Brisbane's floodplain communities were more exposed and vulnerable to flooding than they expected to be. As recounted by Heazle et al. (2013) an essentially political decision had been made in the management of the dam by state government, through the auspices of technical experts in Seqwater and the state Department of Environment and Resource Management (DERM) in 2010 to prioritise water supply over flood risk management in advance of the 2011 floods. This political decision was couched in technical terms – to maintain 100 per cent full supply level – and was left in the hands of technical experts. Notwithstanding the fact that this decision was portrayed by many in the aftermath of the floods as technical incompetence, yet was essentially a political decision (Heazle et al., 2013), it nonetheless seems a reasonable policy prioritisation given the Brisbane region's prolonged drought conditions between 2001 and 2009. Yet this political decision had unexpected (by the public at least) ramifications for the degree of vulnerability and exposure to flooding incurred by communities in the flood plain during the 2010–2011 floods, because it was in direct conflict with the assumptions and political priorities of BCC and their prior assessment of Q100 flood risk.

Given that BCC's priority for spatial planning (in the period 1976–2011) was to optimise development in the floodplain within safe limits (BCC, 2014, 2000; Searle, 2010; Stimson and Taylor, 1998; Heywood, 1986; Simsion et al., 1979) then this normative prioritisation ultimately found expression in the derivation of a Q100 level in the decades prior to 2011. Curiously, the Commission of Inquiry didn't specify the dam operating assumptions made for the council's final Q100 estimates in 2003–2004 and appear to have considered such detail unimportant.[18] However, the commission does reveal that none of these estimates took a precautionary approach as per the calculations made by technical experts contracted to advise council between 1996 and 2000, which had assumed that the dam would be at full supply level. In fact this precautionary (and ultimately validated) approach appears to have been continually rejected by BCC.

Given the apparent motives of BCC for keeping their Q100 estimate as low as possible, and the evidence provided by the Commission of Inquiry's investigation, I argue that BCC have addressed their political priorities through the derivation of expert evidence in a manner suggesting both the politicisation of this evidence and a 'scientised' climate risk management process. BCC's use of objective evidence was deemed a necessary component of robust flood risk management practice in line with the evidence-based mandate (Productivity Commission, 2010) and the associated authority granted in liberal democracies to privileged expert authority (Kitcher, 2011). However, this evidence nonetheless requires important normative decisions between state and local government for its derivation, decisions that appear to have been deliberately influenced to align with contrasting political priorities.

The Commission of Inquiry's investigation ultimately reveals how BCC's political priorities about the advantages (to the local economy and to urban spatial planning) of having a Q100 level as low as possible were in direct conflict with competing priorities of state government to ensure adequate water supply for the southeast Queensland region. As a result of this normative conflict, communities on the floodplain in Brisbane have been living and working under an incorrect assumption of flood risk exposure and vulnerability, as expressed through BCC's planning scheme and its Defined Flood Level of 3.7 metres at the Port Office gauge in Brisbane, an estimate that remained unchanged until 2014, despite the flooding events of 2010–2011 and the conflicting operating decisions for Wivenhoe Dam that prevailed at that time (BCC, 2014; QFCI, 2011). That this normative conflict was masked by technical issues, when in fact they required important political decisions about risk prioritisation, tolerability and the relative exposure of communities to climate extremes speaks to the notion of a scientised policymaking process for climate risk and natural resource management in southeast Queensland.

Conclusion

In the analysis presented in Chapters 4 and 6 I demonstrated that adaptation problems contain inevitable normative components that influence how useful and usable climate science and policy evidence can be for informing policymaking. In the case-studies in this chapter, I have considered the politicisation of policy evidence, not just as an inevitable result of accounting for this contingency as technical experts seek to understand these problems through climate research, modelling and presentation of model outputs (what might be referred to as politicisation-by-process), but politicisation as a deliberate adjustment in technical assessment, framing or conclusions of evidence in accordance with policy players' norms and values (a politicisation-by-agency). This latter form of politicisation occurs, I argue, in order to garner evidence legitimacy (i.e., political acceptability) and to suppress explicit debate. This politicisation in turn can ensure tacit acceptance for particular normative views or political priorities through the use of supposed impartial knowledge, that is, to scientise policymaking in order to garner policy legitimacy.

Risk assessment is often considered a valuable tool for understanding climate adaptation problems. Risk seems an intuitively appealing concept for scientists and policymakers addressing environmental problems since it speaks to the reflexive nature of contemporary problems of modernity in a globalised society (Beck, 1992); it uses a simple empirical formula (Risk = Likelihood X Consequence) that aligns with linear-technocratic norms concerning experts' ability to accurately assess policy problems; and therefore, is seen to allow transparently rational policymaking. However, as demonstrated by the two case-study examples above, risk is also a highly problematic concept when used under a linear-technocratic schema for climate adaptation.

Both likelihood and consequence, and therefore risk itself, resist impartial derivation for climate adaptation problems that are complex, uncertain and subject to divergence of values and priorities. Risk assessment encounters difficulties when used in a way that assumes that it is primarily a technical exercise, that risk can be independently and impartially apprehended, or that any given assessment can encapsulate all (or even most) normative priorities and viewpoints. Technical impact and risk assessment, I argue, can allow norms and values to pass undetected through the assessment process, the results of which will ultimately be used by the political executive as impartial expert evidence.

There are, in summary, a number of factors at play in the processes of politicisation and scientisation described here:

1   The contingent nature of adaptation problems and the trans-scientific, post-normal character of the scientific research available to understand them means

that experts, despite their best and most honest efforts, can never provide definitive or wholly objective assessments of climate hazards or their probability of occurrence.

2   The need for *ex ante* assessment of science for policy as a result of adaptation problem complexity, uncertainty and contingency, suggests that policy evidence must be co-produced between various expert and non-expert policy players in order to ensure that it is sufficiently legitimate and salient to be useful and usable for policy decisions.

3   The prevalence of the linear-technocratic model of policymaking which derives from expectations that experts can apprehend objective reality, that they must do so in order for their evidence to be credible, and that policy can be linearly informed by this impartial scientific and social science research.

These contributing factors raise two important challenges for policymakers when deriving evidence to effectively inform policy decisions.

First, policymakers face a pressing need to balance the legitimacy and salience of policy evidence on the one hand through, for instance, co-production processes, and the credibility of that evidence on the other which, under a linear-technocratic schema, is dependent on the expectation that scientific research and policy evidence should be technical, impartial and provide a measure of objective truth. This need to balance legitimacy, salience and credibility involves important interactions across the science-policy boundary (Cash et al., 2002) and results in tensions between expert and political forms of authority as a result. Even if we renounce all linear-technocratic expectations, it would seem that the science-policy boundary must be both appropriately bridged in order to ensure legitimacy and salience, and sufficiently preserved to ensure its technical credibility.

Second, as a result of this need to balance the competing demands for credibility, legitimacy and salience, policy evidence is in danger of succumbing to processes of politicisation and scientisation. Evidence co-production allows both expert and non-expert policy players to adjust the outputs and conclusions of scientific research in accordance with their preferred norms, values and politics. However, such politicisation may also occur during the commissioning and design of alternative sets of policy evidence, as in the case of the Q100 assessment for the city of Brisbane. Both examples of politicisation were facilitated by the linear-technocratic model. Because of this ability to politicise, I argue, both expert and non-expert policy players are in a position to adjust the content of evidence to suppress further explicit debate or to garner tacit acceptance for their preferred norms and politics under the guise of expert authority, that is, to scientise.

In my examination of these issues, questions arise about whether the *ex ante* policy assessment methods currently used by policy players might be designed in ways that can somehow soothe the ongoing tensions between expertise and politics. If not using concepts of risk, then how else might we conceptualise the hazards presented by climate and climate change to ensure effective adaptation policy? In the next chapter I seek to identify the sorts of evidence that are often identified as missing from climate adaptation policymaking and that can point us in the direction of developing evidence that is both useful and usable in practice. In the final two chapters I will discuss some possibilities for how policy players may effectively address the trade-offs that must be made between credibility, legitimacy and salience, and whether these trade-offs can ever find a sensible or logical resolution.

Notes

1 Weingart (1999) documents a similar scepticism of the role of expertise in policymaking in the US during the mid-to-late twentieth century.
2 Despite their concurrent embrace of 'The Third Way', a social-democratic governance ideal (Giddens, 1998).
3 Relating to a desire for greater autonomy of the market place to manage public goods and services, and a corresponding shrinking of government intervention in matters of public policy (Harvey, 2007: p. 2).
4 As this book is about to go to press, Martin Kowarsch (2016) has published his book *A Pragmatist Orientation for the Social Sciences in Climate Policy*. Kowarsch describes in considerably greater detail the technocratic and decisionist orientations of evidence-based policymaking than I have space to do here, and thus provides a valuable aid in helping to understand the interactions between politics and expertise. For those seeking a more advanced discussion of these issues, therefore, I urge you to read his book.
5 See Chapter 2 for a discussion of the distinction between linear-rational versus technocratic-positivist models of policymaking.
6 Weingart uses the term *scientification* in a very similar way to both Sarewitz' and my own use of the term *scientisation*.
7 So-called 'Reflexive Modernisation' (Beck, 1992).
8 The IPCC's 5 Assessment Report aligns with this definition of risk: 'Risk is often represented as probability of occurrence of hazardous events or trends multiplied by the impacts [or consequences] if these events or trends occur. Risk results from the interaction of vulnerability, exposure and hazard' (IPCC, 2014: p. 1772).
9 This section is adapted from Tangney (2016).
10 The UK Government Department of Environment, Food & Rural Affairs.
11 This section is adapted from Tangney (2015).
12 The Wivenhoe Dam was completed in 1985 and as will be explained here, is a fundamental determinant of flood risk and exposure along the Brisbane River.
13 That is, that the flood mitigation potential of the dams would be at its lowest (see Box 4.4, Chapter 4).
14 Also known as Seqwater, is responsible for the management of the Wivenhoe and Somerset dams.
15 For instance, given the density of urban development along the Brisbane River downstream of the Wivenhoe Dam, it seems reasonable to conclude that any new estimate of Q100 considerably higher than the current value of 3.7 metres would result in significant impacts on property values. Those buildings once designed and built to a safe standard of flood risk protection would no longer be deemed adequate under a revised planning scheme.
16 Curiously, the version of the expert review panel report available online appears to have this value redacted and handwritten in its place '60%' (Mein et al., 2003: 15). It is unknown whether the authors or the online publisher made this amendment; however, neither estimate reflects the dams' actual operating flood mitigation capacity in 2010–2011 (QFCI, 2011: 51).

17   The responsible State Minister and his advisees decided at the time that the Wivenhoe Dam could have released 5 per cent of its full supply capacity immediately in advance of the floods, but they considered this release would have made a negligible enhancement to the dam's flood mitigation capacity (QFCI, 2011: 47).

18   As discussed in above, the 2003 expert review panel report suggests that they were based on assumptions that the dams could reduce downstream peak flood flow rates by about 35 per cent on average (Mein et al., 2003, p. 15).

Bibliography

Banks, G 2009, 'Evidence-based policy-making: What is it? How do we get it?', Chairman's speech, Australian Government Productivity Commission and Australia & New Zealand School of Government ANZSOG/ANU Public Lecture Series, Canberra, Australia, 4 February, viewed 7 November 2014, www.pc.gov.au/speeches/gary-banks/cs20090204/.

Beck, U 1992, *Risk Society: Towards a New Modernity*, Sage, London.

Brisbane City Council [BCC] 2000, *Superseded Brisbane City Plan 2000*, BCC, Brisbane, viewed 5 November 2014, www.brisbane.qld.gov.au/planning-building/planning-guidelines-and-tools/superseded-brisbane-city-plan-2000/.

Brisbane City Council [BCC] 2007, *Brisbane's Plan for Action on Climate Change and Energy 2007*. BCC, Brisbane, viewed 22 July 2014, www.brisbane.qld.gov.au/sites/default/files/20140414%20-%20Brisbanes%20Plan%20for%20Action%20on%20Climate%20Change.pdf.

Brisbane City Council [BCC] 2013, *Flooding in Brisbane: An Explanation of Technical Flood Terms*, Brisbane City Council website, viewed 14 September 2015, www.brisbane.qld.gov.au/community/community-safety/disasters-emergencies/types-disasters/flooding/understand-your-flood-risk/.

Brisbane City Council [BCC] 2014, *Brisbane City Plan 2014 – Schedule 1 Definitions*, BCC, Brisbane, viewed 5 November 2014, http://eplan.brisbane.qld.gov.au/CP/Definitions#DefFloodEvent/.

Cash, D, Clark, W, Alcock, F, Dickson, N, Eckley, N and Jager, J, 2002, *Salience, Credibility, Legitimacy and Boundaries: Linking Research, Assessment and Decision-making*, John F. Kennedy School of Government, Harvard University, Faculty Research Working Papers Series, November 2002.

City Design 1999, *Brisbane River Flood Study*, Brisbane City Council, June 1999, viewed online 1 October 2015, http://resources.news.com.au/files/2011/01/20/1225991/887259-110121-brisbane-flood-study-jun-1999.pdf.

Davies, HTO, Nutley, SM and Smith, PC 1999, 'What works? The role of evidence in public sector policy and practice', *Public Money & Management*, vol. 19, no. 1, pp. 3–5.

Department of Environment, Food & Rural Affairs [DEFRA] 2009, *Adapting to Climate Change: Helping Key Sectors to Adapt to Climate Change*, DEFRA, London, viewed 11 March 2013, http://archive.defra.gov.uk/environment/climate/documents/interim2/report-guidance.pdf.

Department of Environment, Food & Rural Affairs [DEFRA] 2011a, *Greenleaves 3: Guidelines for Environmental Risk Assessment and Management*, DEFRA, London, viewed 17 March 2015, www.defra.gov.uk/publications/files/pb13670-green-leaves-iii-1111071.pdf.

Department of Environment, Food & Rural Affairs [DEFRA] 2012a, *UK Climate Change Risk Assessment: Government Report*, HMSO, Norwich, UK, viewed 6 May 2013, www. defra.gov.uk/publications/files/pb13698-climate-risk-assessment.pdf.

Department of Environment, Food & Rural Affairs [DEFRA] 2012b, *UK Climate Change Risk Assessment: Evidence Report*, DEFRA, viewed 6 May 2013, http://randd.defra.gov. uk/Default.aspx?Menu=Menu&Module=More&Location=None&Completed=0&Pro jectID=15747#RelatedDocuments/.

Douglas, M and Wildavsky, A 1983, *Risk and Culture*, University of California Press, London.

Douglas, H 2009, *Science, Policy, and the Value-Free Ideal*, University of Pittsburgh Press, Pittsburgh, PA.

Fischer, F 2009, 'Technical Knowledge in Public Deliberation: Towards a Constructivist Theory of Contributory Expertise', in Fischer, F 2009, *Democracy and Expertise: Reorienting Policy Inquiry*, Oxford University Press, Oxford.

Giddens, A 1998, *The Third Way: The Renewal of Social Democracy*, Polity Press, Cambridge.

Habermas, J 1971, *Toward a Rational Society*, Heinemann, London.

Harvey, D 2007, *A Brief History of Neoliberalism*, Oxford University Press, Oxford.

Head, B 2008b, 'Three lenses of evidence-based policy', *Australian Journal of Public Administration*, vol. 67, no. 1, pp. 1–11.

Heazle, M, Tangney, P, Burton, P, Howes, M, Grant-Smith, D, Reis, K and Bosomworth, K 2013, 'Mainstreaming climate change adaptation: An incremental approach to disaster risk management in Australia', *Environmental Science & Policy*, vol. 33, pp. 162–170.

Hess, M and Adams, D 2002, 'Knowing and skilling in contemporary public administration', *Australian Journal of Public Administration*, vol. 61, no. 4, pp. 68–79.

Heywood, P 1986, 'Brisbane's development in zones: A review of the 1986 city of Brisbane town plan', *Australian Planner*, vol. 24, no. 4, pp. 28–33.

Holmes, J and Clark, R 2008, 'Enhancing the use of science in environmental policy-making and regulation', *Environmental Science & Policy*, vol. 11, pp. 702–711.

Hoppe, R 2005, 'Rethinking the science-policy nexus: From knowledge utilization and science technology studies to types of boundary arrangements', *Poiesis Prax*, vol. 3, pp. 199–215.

Intergovernmental Panel on Climate Change [IPCC] 2014, 'Summary for policymakers', in Field, CB, Barros, VR, Dokken, DJ, Mach, KJ, Mastrandea, MD, Bilir, TE, Chatterjee, M, Ebi, KL, Estrada, YO, Genova, RC, Girma, B, Kissel, ES, Levy, AN, MacCracken, S, Mastrandea, PR, White, LL (eds) 2014, *Climate Change 2014: Impacts, Adaptation and Vulnerability. Part A: Global and Sectoral Aspects. Contribution of Working Group II to the Fifth Assessment Report of the Intergovernmental Panel on Climate Change*. Cambridge University Press, Cambridge and New York.

Jasanoff, S 1990, *The Fifth Branch: Science Advisers as Policymakers*, Harvard University Press, Cambridge, MA.

Jasanoff, S 2003a, '(No?) Accounting for expertise'. *Science and Public Policy*, vol. 30, no. 3, pp. 157–162.

Jasanoff, S 2005, 'Judgement under Siege: The Three-Body Problem of Expert Legitimacy', in Maasen, S and Weingart, P (eds), *Democratization of Expertise? Exploring Novel Forms of Scientific Advice in Political Decision-Making*, Springer, Dordrecht, the Netherlands.

Kay, A 2011, 'Evidence-based policymaking: The elusive search for rational public administration', *The Australian Journal of Public Administration*, vol. 70, no. 3, pp. 236–245.

Kitcher, P 2011, *Science in a Democratic Society*, Prometheus Books, New York.

Kowarsch, M 2016, *A Pragmatist Orientation for the Social Sciences of Climate Policy: How to Make Integrated Economic Assessments Serve Society*, Springer International Publishing, Basel.

Lash, S and Wynne, B 1992, 'Introduction', in Beck, U 1992, *Risk Society: Towards a New Modernity*, Sage, London.

Maasen, S and Weingart, P 2005, 'What's New in Scientific Advice to Politics?', in Maasen, S and Weingart, P (eds), *Democratization of Expertise? Exploring Novel Forms of Scientific Advice in Political Decision-Making*, Springer, Dordrecht, the Netherlands.

Majone, G 1989, *Evidence, Argument and Persuasion in the Policy Process*, Yale University Press, New Haven, CT, and London.

Marston, G and Watts, R 2003, 'Tampering with the evidence: A critical appraisal of evidence-based policymaking', *The Drawing Board: An Australian Review of Public Affairs*, vol. 3, no. 3, pp. 143–163.

Mein, R, Apelt, C, Macintosh, J and Weinmann, E 2003, 'Review of Brisbane River Flood Study: Report to Brisbane City Council', Brisbane City Council, September 2003, viewed online 1 October 2015, http://resources.news.com.au/files/2011/01/20/1225991/885887-110121-flood-studysep-2003.pdf.

Nieman, M and Stambough, SJ 1998, 'Rational choice theory and the evaluation of public policy', *Policy Studies Journal*, vol. 26, pp. 449–465.

Nutley, S, Davies, H and Walter, I 2002, *Evidence Based Policy and Practice: Cross Sector Lessons from the UK – Working Paper 9*, ESRC UK Centre for Evidence Based Policy and Practice, Research Unit for Research Utilisation, Department of Management, University of St Andrews, Edinburgh.

Oreskes, N 2004, 'Science and public policy: What's proof got to do with it?', *Environmental Science and Policy*, vol. 7, pp. 369–383.

Productivity Commission 2010, *Strengthening Evidence Based Policy in the Australian Federation*, Volume 1: *Proceedings*, Roundtable Proceedings, Productivity Commission, Canberra.

Queensland Floods Commission of Inquiry [QFCI] 2011, *Queensland Floods Commission of Inquiry: Interim Report*, QFCI, Brisbane, viewed 3 April 2013, www.floodcommission.qld.gov.au/publications/interimreport/.

Queensland Floods Commission of Inquiry [QFCI] 2012, *Queensland Floods Commission of Inquiry: Final Report*, QFCI, Brisbane, viewed 3 April 2012, www.floodcommission.qld.gov.au/__data/assets/pdf_file/0007/11698/QFCI-Final-Report-March-2012.pdf.

Queensland Police 2011, *Media Release: Death Toll from Queensland Floods*, 24 January, viewed 12 December 2012, www.police.qld.gov.au/News+and+Alerts/Media+Releases/2011/01/death_toll_jan24.htm/.

Renn, O 2008, *Risk Governance: Coping with Uncertainty in a Complex World*, Earthscan, London .

Rittel, HW and Webber, MM 1973, 'Dilemmas in a general theory of planning', *Policy Sciences*, vol. 4, no. 2, pp. 155–169.

Sanderson, I 2002, 'Evaluation, policy learning and evidence-based policymaking', *Public Administration*, vol. 80, no. 1, pp. 1–22.

Sarewitz, D 2004, 'How science makes environmental controversies worse', *Environmental Science & Policy*, vol. 7, pp. 385–403.

Searle, G 2010, 'Too concentrated? The planning distribution of residential density in SEQ', *Australian Planner*, vol. 47, no. 3, pp. 135–141.

Seqwater 2015, *Water for Life: South East Queensland's Water Security Program 2015–2045*, Seqwater, July 2015, viewed online 26 December 2016, https://www.google.com.au/url?sa=t&rct=j&q=&esrc=s&source=web&cd=1&ved=0ahUKEwiHoemU95DRAhV

Kp5QKHR1NBOUQFggeMAA&url=http%3A%2F%2Fwww.seqwater.com.au%2Fsit
es%2Fdefault%2Ffiles%2FPDF%2520Documents%2FWater%2520for%2520life_
Water%2520Security%2520Program.pdf&usg=AFQjCNFIpOMuLUq39moN-QXlCe
5Bzm1NTQ&sig2=RYP4VQFRcAlW321L01UduA&cad=rja/.

Simsion, DW, Kerry, MJ, Llewellyn-Smith, MJ, Day, P, Borrows, EA, Graham, L and Hevgold, L 1979, 'Seminar B - Metropolitan planning - Adelaide, Brisbane, Perth', *Royal Australian Planning Institute Journal*, vol. 17, no. 1, pp. 97–107.

Solesbury, W 2001, 'Evidence Based Policy: Whence it came and where it's going – Working Paper 1', ESCRC UK Centre for Evidence Based Policy and Practice, Queen Mary, University of London.

Stimson, R and Taylor, S 1998, 'Dynamics of Brisbane's inner city suburbs', *Australian Planner*, vol. 35, no. 4, pp. 205–214.

Tangney, P 2015, 'Brisbane City Council's Q100 assessment: How climate risk management becomes scientised', *International Journal of Disaster Risk Reduction*, vol. 14, no. 4, pp. 496–503.

Tangney, P 2016, 'The UK's 2012 climate change risk assessment: How the rational assessment of science develops policy-based evidence', *Science and Public Policy*, published online 6 September 2016, https://academic.oup.com/spp/article-abstract/doi/10.1093/scipol/scw055/2525558/The-UK-s-2012-Climate-Change-Risk-Assessment-How.

Weingart, P 1999, 'Scientific expertise and political accountability: Paradoxes of science in politics', *Science and Public Policy*, vol. 26, no. 3, pp. 151–161.

Willows, R and Connell, R (eds) 2003, *Climate Adaptation: Risk, Uncertainty and Decision-Making*, UK Climate Impacts Programme, Oxford.

8 Evidence needs for adaptation policymaking

> Science not only has a methodology, but it also has a history, a geography and a sociology.
>
> Mike Hulme

In this chapter I seek to understand the evidence needs of policy players engaged in climate adaptation and to discuss an alternative conceptual framework for the development of ex *ante* policy evidence. I outline a simple typology for categorising adaptation issues in a way that highlights some of the important variables involved in understanding policy problems. I then investigate a resilience-based framework of policy evidence development as a means to reduce the worst excesses of the prevailing linear-technocratic heuristic and that, I argue, provides an intriguing alternative to risk-based methods (see Chapter 7).

Although it may surprise the reader given my critique of linear-technocracy up to this point, the typology I use here describes the uncertainties relating to adaptation in a traditionally linear way. I do so for a few reasons. First, the typology aligns with the commonplace perspectives of many policy players for whom a progression of *evidence-problem-evidence-solution* has intuitive appeal. The typology is attempting, therefore, to help those accustomed to thinking in this way to meaningfully understand the uncertainties of decision-making for adaptation policy. I use a linear format to show as simply as possible how adaptation problems invoke policy players' evidence needs in different ways depending on what types of uncertainty are most prevalent and where complexity exists in relation to priorities, problems and responses.

Importantly, however, while the typology may help readers to understand and address climate-related policy problems, it should not be viewed as an adequate characterisation of those problems as they are encountered during policymaking per se. As I have stressed in previous chapters, adaptation problems resist definitive synoptic characterisation from any given governance level or scale in a way that is universally applicable, and policy responses often develop concurrently to characterisations of those problems or the development and maturation of specific goals and objectives within policy subsystems.

Nonetheless, this typology may help to further explain the tensions, discussed in Chapter 5, between expertise and politics that arise when seeking to achieve sufficient legitimacy and/or salience of the available evidence for policymaking. Although experts focus primarily on uncertainties concerning the physical and social bounds of impacts associated with climate variability and change, as this typology shows, non-expert policy players often have a broader portfolio of uncertainties and priorities to consider. This disjunction of priorities is another reason for tension at the science-policy interface.

Kirchoff et al. (2013) suggest that the choice and use of evidence sought by adaptation decision-makers is principally determined by their perceptions of climate risk. In turn, they argue, risk constructions are determined by past experiences, the decision-making context and the cognitive processes allowed or preferred in that context. When considering evidence needs for adaptation policymaking, I argue, these determinants are themselves heavily circumscribed by the political executive and by the prevailing norms and prescriptions of the evidence-based mandate. Perceptions of risk, or more appropriately, politically legitimate interpretations of risk, largely depend on the mandate of prevailing politics upon the policymaking community. Cognitive processes meanwhile are, if not largely prescribed, then certainly constrained by the imperative to use rationalist methods of *ex ante* policy analysis advocated by many liberal democracies, including the two cases in question for this research (see for example, NEMC, 2010; Willows and Connell, 2003). Even without such prescriptions, and notwithstanding bureaucratic politics, policymaking seems generally more amenable to careful considered analysis of evidence, than intuitive, experiential responses.[1]

Nonetheless, I argue, although linear-rationalist assumptions about what evidence is and can do may be entrenched, policy players need not persist with risk-based methods that are fundamentally ill-suited to the task at hand. In this vein, therefore, I explicate the concept of climate resilience and its usefulness and usability for adaptation problems. Although not without its own difficulties, resilience-based decision-making, I believe, aligns more appropriately with the characteristics of adaptation policy problems and the uncertainties and information requirements for policy players attempting to resolve them. To begin, however, I discuss the perceived evidence needs of experts for enhancing adaptation decision-making.

Perceived evidence needs for climate adaptation

As Kirchoff et al. (2013) and Preston et al. (2013) suggest, adaptation decision-makers perceive that there is a deficit of adequate information for adaptation policymaking and practice in developed countries. This view sits alongside a growing criticism of the information that is already available. Preston et al. (2013) argue that those critiques suggest that adaptation science is growing without adequate understanding of how it can or should contribute to effective adaptation. As discussed in previous chapters, in the face of climate change's wicked characteristics; the ongoing and potentially increasing frequency of extreme events; and the lack of clearly articulated policy goals and objectives, it would seem that adaptation options often develop concurrently with the development and use of expert evidence, not as a result of it. This conclusion suggests a need to reconsider long-held expectations for how and what evidence and expertise can provide for policy decision-making.

Notwithstanding the difficulties of adequately understanding adaptation problems, the principal focus of governments' adaptation plans to date has been on the development of more and better evidence to linearly inform policymaking and, in particular, on understanding climate risks on the basis of projections of future climate change. As described in Chapter 7, adaptation plans are often accompanied by assumptions that climate change projections can, or will eventually, provide decision-relevant predictions that will allow policy players to apprehend adaptation problems in a relatively definitive way, and on that basis, to optimise adaptation responses to the most likely future outcomes. This prescription appears to be fuelled, at least in part, by the scientific community (see for example, Pitman, 2016; Church et al., 2016). As discussed in Chapters 4 and 6, however, such assumptions are unrealistic and ill-advised in practice. Not only are climate predictions not possible, adaptation problems otherwise resist definitive characterisation in ways which suggest that optimised adaptation solutions can result in maladaptation. Moreover, due to these uncertainties, policymakers are unlikely to assent to disruptive adaptation options, policy pathways or significant capital investments that are optimised to any scenario other than business-as-usual (see Chapter 3).

While the adaptation policymaking community has all too often assumed a linear-rationalist 'predict-then-act' approach to climate adaptation, there has also been a relative absence of expression of underlying goals by policymakers in relation to the resilience of society and the economy. And rather surprisingly, adaptation plans have also rarely specified tangible objectives in relation to the adaptation and resilience of particular jurisdictions, risk-receptors or government portfolios to climate change that might direct the development of effective evidence, other than bland statements relating to maintaining the structure and function of existing socio-economic and ecological arrangements. This may also be one of the contributing factors to accusations of a continuing dearth of actual policy action for climate change adaptation, irrespective of the presence of a considerable number of plans and programmes (see Chapter 3). Indeed, as Grundmann's (2006) study of climate politics in the US has demonstrated, the promise of more and better evidence can be a useful compromise between partisan political views on climate change. This is because, in the eyes of those opposing climate change policies, the foreseeable result of more and better evidence is a near-term lack of unpalatable policy action, while the rationalist expectations of those who advocate climate change action also necessitate more and better evidence.

The potential lack of immediate clarity of goals and objectives in relation to climate adaptation, and the need for user input into identifying research focus and evidence development are issues that highlight the importance of understanding co-production processes and the possibility of political influence in the development of adaptation science and policy evidence (Preston et al., 2013). In the absence of clear objectives, policy players may seek to influence evidence development as a means of influencing whatever policy actions may follow in line with their values and priorities. Thus, Meyer (2011) advocates for more open acknowledgement of the values and priorities implicit in both research funding and in research itself, given the uncertainties, contingencies and politics of the available knowledge.

If understanding adaptation problems often runs concurrent to managing those problems this indicates that both our understanding of and response to adaptation problems may depend quite significantly on the normative preferences of policy players and the consensus they are able to achieve during the co-production of policy evidence. This contingent character of evidence for climate adaptation highlights the importance of making explicit (even if only tentative) goals and priorities about adaptation and resilience during the development of policy evidence, as well as the dangers of scientisation processes that would seek to prescribe and legitimise policy responses on the basis of supposedly objective, impartial and adequate technical knowledge.

A typology of uncertainty for adaptation policy problems

As described in Chapters 3 and 7, prevailing stable and dynamic political factors can have a significant bearing on what types of evidence are prepared and used for policymaking. In Australia, climate *change* science has largely been eschewed for state and local government climate adaptation planning in favour of economic evidence demonstrating a viable business case for policy action. In the UK, under the remit of the Climate Change Act (2008), there has been an expectation by government that the best available climate change projections and adaptation science should inform adaptation policy, while outside this remit, local government has returned to a 'business-case' model and increasingly uses the concept of climate resilience, perhaps as a means of promulgating neo-liberal values (Welsh, 2014). Alongside prevailing political priorities, however, these comparative circumstances can also demonstrate how the presence and locus of decision-making uncertainty can play a decisive role in understanding evidence needs and policy responses for climate adaptation.

For the purposes of demonstrating the evidence needs of policy players, I propose that uncertainties relating to the management of adaptation problems may broadly fall into three categories:

1   **Uncertainty of policy goals and objectives** – what government wants to achieve and why.
2   **Uncertainty of adaptation problems** – why and how are adaptation issues a problem and how can government understand the associated hazards and opportunities.

3 **Uncertainties relating to policy responses** – how can and should government respond to these climate problems and what are the implications of any proposed response?

Interestingly, there appear to be very few strategic-level adaptation policies, plans or evidence-based assessments across the two case-study regions that systematically address all three of these types of uncertainty. Perhaps, as I argue further below, they are not fully addressed because of the nature of the adaptation policy challenge and the linear-technocratic ideals that circumscribe its attempted resolution.

Adaptation at a most basic level is a response to a stimulus, and adaptation policy, therefore, relates to the adjustment, removal or replacement of existing policies or policy objectives to ensure the achievement of underlying priorities. Given the potentially contentious nature of such a task, it is unsurprising that many climate risk assessments and adaptation plans fail to state for which underlying objectives and goals risks are actually being assessed and managed. The assumption appears to be that these relate to a well-accepted status quo, to explicit but nonetheless under-specified priorities of government such as the pursuit of economic growth, or to the preservation and stability of existing socio-economic systems and the communities they serve (see for example, COAG, 2007; 2011). There seems little reason, on the face of it, to be any more specific about objectives and priorities.

Unsurprisingly, therefore, most strategic climate risk assessments and guidelines examined for this book concentrate their attention on pursuing a formula to quantify or otherwise systematise our understanding of climate hazards and impacts, and on that basis decide about what government should do to address them; meanwhile, underlying values and priorities remain largely implied or hidden (see for example, Queensland Government 2012b; COAG, 2007). As I argue here, however, these underlying values and priorities may often relate to interpretations of climate vulnerability and *resilience* that are fundamentally important for how we formulate the task of adaptation and tell us much about how we view the nature of our communities and the economy.

The climate change risk assessment conducted by the Environment Agency of England & Wales in 2010, was a partial exception to this trend. Although this assessment is limited by its focus on the Agency's corporate objectives, such is the national remit of its climate adaptation policy implementation and regulation in the UK, it can suffice as a proxy for how strategic government assessments might proceed if such priorities were sufficiently tangible and clarified. I will use this example to demonstrate the utility of the typology presented here for identifying information requirements, and how the three types of uncertainty described above may combine and relate in the practice of adaptation policymaking.

The Environment Agency's adaptation report seems a useful case-study for this typology, given that it is the largest non-departmental government body responsible for managing the natural environment in the UK (indeed, the largest of its kind in Europe) (EA, 2010a). Not only had the Agency clear and explicit strategic goals at the time of the assessment that it was obliged to follow, and a long tradition of pursuing evidence-based approaches, it also had wide-ranging responsibilities relating to:

- flood risk and coastal erosion management;
- water resources and water quality management;
- the environmental regulation of industry;
- land quality;
- wildlife and habitats;
- riparian navigation and recreation; and
- sustainable urban and rural planning guidance.

In 2009, the Environment Agency was asked by the UK Government to report its risks and opportunities from climate change and its plans to address them, as a requirement under the so-called 'Adaptation Reporting Power' of the Climate Change Act (2008) (DEFRA, 2009). Given its extensive responsibilities, the Environment Agency was asked to provide a report that would serve as a guide to other statutory authorities, infrastructure and service providers with reporting obligations under the Act[2] (EA, 2010a).

Alongside this example from the UK, I will also use three examples of adaptation problems from southeast Queensland as a comparative case to demonstrate the typology's generalisability. The underlying goals and priorities relating to southeast Queensland's adaptation problems are much less well described in policy literature, or at least not as unambiguously as for the Environment Agency's risk assessment, and as such, the typology makes an assumption concerning the relative certainty associated with state and local governments' risk-specific policy objectives, based on broader policies relating to state and federal emergency and disaster risk management, and community disaster resilience (Queensland Government, 2012b; COAG, 2011).

Using the uncertainty criteria above, I demonstrate four indicative types of adaptation problem, with reference to the Environment Agency's responsibilities listed above for southeast England (SEE), alongside three of the most damaging and costly climate risks to the southeast Queensland (SEQ) region: flooding, drought and storm/cyclone wind damage (Risk Frontiers, 2012). These risks and their categorisation using these uncertainty criteria are outlined in Table 8.1.

Type 1 Adaptation problems

Type 1 problems are those for which there is general consensus (at least at face value) about the nature of the problem, what government seeks to achieve in addressing it, and how it may be best resolved. For the issue of both inland and coastal flooding in the UK, for instance, although individual communities may perceive the associated risks differently depending on their particular location

Table 8.1   A typology of uncertainty for common adaptation policy problems

Type	Policy objectives	Risks and opportunities	Policy responses	Example
1	Low uncertainty: Clear adaptation objectives	Low uncertainty: Few competing risks	Low uncertainty: Established options	**SEE:** Flood risk and coastal management **SEQ:** Cyclones and storm management
2	Low uncertainty: Clear adaptation objectives	Medium uncertainty: Many competing risks	Medium uncertainty: Competing options	**SEE:** Water resources and quality management; industry regulation; riparian management and infrastructure provision
3	High uncertainty: Uncertain or ambiguous adaptation objectives	High uncertainty: Many competing and uncertain risks	High uncertainty: Few well-established options, many possibilities	**SEE:** Wildlife and habitat management
4	High uncertainty: Ambiguous adaptation objectives	Medium uncertainty: Continually competing risks	Medium uncertainty: Competing options in a zero-sum game of risk management	**SEQ:** Flood risk and water resources management – (operation of the Wivenhoe and Somerset dams – see Box 4.3)

and circumstance, government and its arms-length bodies such as the Environment Agency have long pursued a strategy to manage these combined risks and vulnerabilities in a coherent and cost-effective way. Broadly speaking, this strategy seeks to ensure that risk is minimised for those communities already situated in exposed locations and that climate vulnerability should be minimised in the future by minimising the amount of new development in flood risk-prone areas (EA, 2010a). There are exceptions to this policy that highlight the contentious and context-specific nature of adaptation problems (see Box 4.1, Chapter 4), but for the most part government has a set of strategic policy objectives that are coherent at the level of national governance. These objectives exist irrespective of climate change.

Furthermore, the risks from climate change (sea-level rise and the increasing intensity and frequency of fluvial and coastal flooding events) are an exacerbation of existing risks associated with climate variability and therefore do not present new forms of risk that are likely to compete for the Environment Agency's attention *within the flood risk management portfolio*. Thus, policy responses such as infrastructure or planning provisions that address extant climate vulnerabilities and pressures involve the same types of infrastructure and planning provisions that are expected to be needed to minimise the risks from climate change.

Established options for the management of flood risk are unlikely to be replaced with alternative risk management options since so few alternatives exist. That is not to suggest that the Environment Agency fully understands those risks or how effective their adaptation responses will ultimately be in the face of climate change, but that the strategy and tactics for addressing these risks are unlikely to change.

In this sense, there may well be a degree of path dependence associated with responses to Type 1 problems which means that poor pre-emptive decisions in relation to infrastructure provision and planning responses are likely to be costly to both communities and government. The established options are also often subject to cost-benefit assessment that discount the future over the present, making some risks, such as those faced by the residents of Happisburgh (see Box 4.1, Chapter 4), less amenable to capital investment than others because the costs of infrastructure provision outweigh the benefits to one small community.

Likewise, for the case of cyclone and storm management in southeast Queensland, policy objectives, although often not explicitly stated, seem to relate to maintaining the stability of existing communities and their socio-economic systems (see for instance, COAG, 2011: p. 5; NEMC, 2010: p. 4) which refer to the concepts of self-reliance, social capacity and the creation of 'safer, more sustainable communities'). In the context of addressing the risks from extreme winds, Australian federal government suggests that there are few competing risks (Australian Government, 2007: p. 16–21). Flooding is the principal concurrent risk, the dynamics of which are relatively well understood and is dealt with concurrently by emergency services and the communities they serve in the aftermath of a storm or cyclone event. Meanwhile, established preparation and response options may relate to climate-proofed design of public buildings and amenities, public information and engagement programmes, co-produced community preparedness plans and effective emergency response through local, district and state disaster management groups (Queensland Government, 2012b; Australian Government, 2007; EMA, 2004). Although there is considerable uncertainty about the future risks from climate change in relation to cyclone and storm frequency and intensity, it is generally expected that, with refinement and sophistication of public engagement initiatives (see for example Howes et al., 2015), existing established responses will continue to be relied upon (Australian Government, 2007).

Type 2 Adaptation problems

Type 2 problems are also subject to relative consensus and certainty in relation to policy objectives; however, uncertainty lies in understanding and prioritising risks and in government's available options to manage them. For the case of water management or industry regulation in the UK, for instance, objectives have been established through regulatory or legislative prescriptions for the management of environmental hazards that seek to ensure that the public have access to clean, abundant water supplies and an unpolluted landscape. In the UK case, many of these legislative prescriptions have, until recently, come from the EU and thus national government have been obliged to pursue certain environmental quality standards, even though the manner of the underlying legislation allows some freedom of interpretation in its implementation. However, climate risks can be challenging to understand for Type 2 problems and the spectre of climate change complicates our understanding of extant and concurrent environmental risks to water resources and quality, and from industry.

There is, for example, considerable uncertainty concerning what the net effect of warming fresh waters will have upon water quality already threatened by increasing urbanisation, over-abstraction of water resources and intensive agricultural practices in the UK (EA, 2010a). Climate change also challenges the means and extent to which existing policy prescriptions and objectives may be achieved, in ways that seem considerably more complex than Type 1 adaptation problems. This is either because these legislative and policy provisions may be considerably less achievable under a changed climate regime (e.g., in the case of existing water quality objectives), or because viable adaptation policies have not yet been devised that can counter the potential impacts from climate change, and so there is some uncertainty regarding the best available response options (as, for instance, in the case of water resources provision in the UK) (EA, 2010a). Unlike Type 1, the best options available for Type 2 problems may not be the ones that have conventionally been relied upon in the past. For example, climate change may result in a dependence on desalination plants or increasing reliance on regional transfers of water resources, rather than the UK's traditional reliance on water abstraction licensing and reservoirs.

Type 3 Adaptation problems

Type 3 adaptation problems are those for which uncertainty is greatest. For the case-studies used here they are principally relevant to the management of species, habitats and ecosystems. These types of adaptation problems are subject to intractable uncertainties relating to the viability of objectives, our limited knowledge of both future climate change and concurrent risks associated with social-ecological management, and therefore also our understanding of how best to manage those risks. In the UK case, the policy objectives associated with Type 3 problems have often come from EU legislation and relate to the conservation of habitats and ecosystems and the preservation of particular species that may be endangered, that have significant economic value attached to them, or both. However, the objectives of habitat conservation and species preservation may not necessarily be compatible in the presence of a changing climate or the effects of concurrent anthropogenic pressures such as increasing urbanisation.

For example, the species-specific objectives of EU legislation in relation to the preservation of *Salmonid* fish species may not be achievable if climate change results in conditions that are not conducive to their preservation (i.e., there are limited contingencies for what to do if salmon or trout simply cannot survive in warmer waters in UK streams and rivers as a result of climate change). Furthermore, it is possible that some habitat conservation tactics in the face of climate change may be helpful for a given habitat as a whole, but be detrimental to the preservation of particular species. These difficulties in agreeing policy objectives and responses are compounded by considerable uncertainties associated with the dynamics and functioning of social-ecological systems which mean that it can be very difficult to understand the potential effects from climate change.

Type 4 Adaptation problems

Finally, *Type 4* describes a type of adaptation problem which can result from the path-dependency or complexity arising from past decisions. This type of adaptation problem is currently relevant for the flood and drought risk management strategies of southeast Queensland, and which highlights the dangers of rationalist 'predict-then-act' approaches to adaptation policy. Such an approach can result in policy options that have become wholly path-dependent and constrained by the types of policy decisions made by past decision-makers. As described in Box 4.4, Chapter 4 the Wivenhoe and Somerset dams in southeast Queensland manage both water resources and flood risk. As such, two climate risks and adaptation goals are in perpetual conflict as a result of the dual purposes for which these dams were designed and the resulting zero-sum game between their opposing operational modes. In the face of climate change, these management difficulties may become even more acute. Type 4 adaptation problems, therefore, are those for which past policy decisions have tied policymakers into a series of pre-determined options on the basis of either legislative constraints or physical infrastructure.

Deriving evidence needs from political and scientific uncertainties

The four types of adaptation problem outlined above are indicative examples, but should not be taken as a comprehensive typology, less still, an adequate theoretical characterisation. Moreover, the uncertainties associated with each of the problems outlined above across the three categories may change as experts learn more about climate vulnerabilities and adaptation responses, or their attitudes to climate risk and vulnerability change over time.[3] Even so, the categories of potential uncertainty are unlikely to change and remain relevant irrespective of the available evidence. They provide a useful foundation for understanding the knowledge requirements for adaptation policymaking.

Importantly, however, the underpinning criteria should be used with some care. This typology must be considered in the context of the non-linearity of policymaking. Issue identification, understanding, objective-setting and policy responses often run concurrently rather than consecutively, contrary to what is often suggested by linear-technocratic schema. These limitations have important implications for how we seek to understand these problems and how we commission and use expert evidence. This typology's ability to account for the wickedness of adaptation problems is also constrained by the contingency inherent in this characterisation; it accounts for adaptation problems at a strategic governance level that may only be relevant for certain policy players. So, for instance, the uncertainties and conflicts described for the Environment Agency's climate change risks above are specific to that institution, even though the problems described are relevant to many different groups of stakeholders. The intractability of these problems, however, may become even more pronounced when considered across governance levels and scales. As described in Chapters 4 and 6, the consensus achieved at one particular governance level concerning the characteristics of an adaptation problem may not necessarily exist across them.

Although there may be consensus of objectives between the Environment Agency and the UK government in such a way that is congruent with EU legislative requirements, there may still be conflicts between these objectives and those of local authorities or communities who may feel disenfranchised by objectives and policy responses that have been determined at national or international levels and that fail to consider their specific needs and circumstances. The case of the village of Happisburgh described in Box 4.1, Chapter 4 is a useful case in point. In the context of such difficulties of cross-level governance, as well as the considerable uncertainties associated with mono-scalar assessment and management of climate risks, it seems reasonable to suggest that political or normative considerations are a necessary and important constituent of evidence development. These considerations become important when determining the nature of adaptation problems, what counts as legitimate evidence and expertise, and when determining adaptation policies and their underlying objectives.

The typology outlined above suggests that certain evidence sources are more useful for some types of adaptation problem than for others. For example, where objectives and response options are well understood and relate to the elimination of hazard exposure or vulnerability through the establishment, upgrading or maintenance of infrastructure or engineering provisions – as in many Type 1 and Type 2 problems – adaptation responses may be derived with the use of climate change projections through a process of exploratory (not predictive!) scenario planning (Dessai and Hulme, 2004). This is not to suggest that pre-emptive investment in climate change adaptations will pass neo-liberal governments' cost-benefit test. This seems unlikely. Nor do these uses of climate science avoid the difficulties of uncertainty and the mismatch of scale between this science and individual policy problems. These difficulties preclude the sort of optimised adaptation responses sometimes envisaged by the climate science community. However, as described for the case of the Thames Estuary 2100 project in southeast England (see Box 4.3, Chapter 4), climate change projections can be used to inform the progressive development of adaptation responses that are designed to be robust to a broad range of potential climate outcomes as well as to existing climate variability and extremes. An exploratory planning approach would seek to understand potential changes to a limited number of key variables relating to the tolerances and thresholds of infrastructure and engineering to climate variability and change.

Alternatively, where objectives, risks and response options are less clearly defined, understood or they are more contested due to the variety of impact and response variables at play and the many associated uncertainties, as in Type 3 and Type 4 problems, climate change projections appear to be much less useful. For such problems, evidence needs may be directed more toward understanding the specific vulnerabilities and resilience of individual risk-receptors, locations or communities – so-called 'bottom-up' approaches (Dessai and Hulme, 2004). Further, in such circumstances where policy objectives are considerably less clear, or ambiguous, as for Type 3 and Type 4 problems, it may not be possible to understand precisely how to respond without a better understanding of the thresholds and tipping points inherent in social-ecological systems (Moss et al., 2013) or the political priorities of government (Heazle et al., 2013), irrespective of how good the available climate science may be.

Importantly, the identification of evidence needs using this typology is dependent on input from decision-makers and other stakeholders and should not be a task that is assumed to be the sole responsibility of the expert community. Our norms, ideals and values relating to climate risk may determine much about how we choose and use evidence and derive adaptation policy. Adaptation policymaking often incorporates, for example, policy players' implicit conceptions of resilience. It is to this idea of climate resilience that I now turn.

Conceptions of resilience for climate adaptation policymaking

In previous chapters I outlined the limitations of risk-based approaches to policy evidence development, whereby, the strict empirical or semi-empirical assessment of likelihood and consequence is problematic in practice and can even foment the politicisation of expert authority for decision-making under a linear-technocratic schema. Risk, in the hands of policymakers and their chosen experts, too often assumes an ability to definitively account for climate effects on social-ecological systems; it allows policymakers to avoid transparent expression of values and priorities in relation to what and how we know about the climate through their recourse to supposedly independent and objective advice. Here I briefly explore the possibility for climate resilience as the foundation for an alternative framework for the derivation of policy evidence.

Resilience is a concept that is relevant to all governance and organising activities concerned with the viability of human settlements and the communities they sustain. Its use as a policymaking framework necessitates a range of (often implied) positions concerning human interactions with the natural environment: the functioning of social-ecological and economic systems, prevailing values in relation to social conservatism, economic (neo-)liberalism and liberal democratic governance, as well as, potentially, government's attitudes to scientific evidence and expert judgement on climate change. If appropriately operationalised as a *reflexive epistemic framework*, I believe it shows considerable potential as an alternative means of framing scientific evidence; one that can better inform policy players' and the public about the normative positions and political priorities of government that underpin adaptation objectives, risks and management options, and therefore can account for each of the uncertainty categories described through the typology above.

Resilience can be used in a variety of ways to elucidate the strengths of individuals, communities and/or their ecological support systems. Unlike traditional risk-based approaches, however, resilience does not assume an ability to reduce pluralistic policy problems to *scientistic* terms that are ultimately dependent on the understanding provided through necessarily limited perspectives, or that are limited by the intractable uncertainties of future unknowable events (Chandler, 2014). Understanding and seeking to enhance climate resilience can nonetheless incorporate issues of climate risk, vulnerability and exposure in ways that, I argue, may circumvent the inadequacies (i.e., a lack of salience and legitimacy) of existing climate science outputs, while adequately addressing political and economic norms and values in a way that enhances the usefulness and usability of policy evidence.

Whether intended or not, interpretations of resilience are already often implicit (if not also under-specified) in experts' formulations of climate change risk, adaptation problems, and in policymaking strategies and practices, and therefore, the concept holds considerable importance as a foundation for our understanding and governance of climate adaptation. As Prosser and Peters (2010) point out, although differences in the interpretation of resilience may at first appear trivial, they can result in significant differences to policy objectives, capabilities and costs, and by implication, policy outcomes and their efficacy. Indeed, when used as a normative prescription for climate adaptation, interpretations of resilience may mask a variety of political and socio–economic ideals and priorities (Hornborg, 2013). But is resilience any better at masking policymakers' values and priorities than the prevailing climate risk management paradigm for the development and interpretation of policy evidence? Here I suggest that it is considerably less effective in this regard and perhaps therein lies its potential when developing credible, salient and legitimate policy evidence. Interpretations of resilience can tell us much about governments' political ideologies and priorities (Cretney; 2014; Davoudi, 2012), in ways that risk-based approaches appear to suppress or ignore.

Contemporary concepts of resilience have been investigated or used across the academic disciplines of ecology, psychology, human geography, urban planning, economics and disaster risk management, amongst others (Davidson et al., 2016; Welsh, 2014; Adger et al., 2011; Carpenter et al., 2005; Funfgeld and McEvoy, 2012; Kuhlicke et al., 2011; Maguire and Hagan, 2007; Walker et al., 2004; Yohe and Tol, 2002). Despite various interpretations of the concept, all literatures are similarly concerned with an ability to cope with stresses and respond effectively to change, whether that is for individuals, species, communities, institutions, infrastructure or ecosystems. Recent literature has brought significant conceptual advancement to the term, attempting to reconcile varying disciplinary interpretations and to resolve previous criticisms of the concept, to allow some practical utility for environmental and social policy, planning and governance (Davidson et al., 2016; Davoudi, 2012; Duit et al., 2010; Pike et al., 2010; Simmie and Martin, 2010). Recent literature focuses on resilience as a heterogeneous normative concept relevant to all efforts to understand how communities – and the SESs of which they are a part – can and should respond to stresses and adapt to and prepare for change. This heterogeneity of interpretations is sometimes highlighted as a limitation of resilience as an operational concept. I argue, however, that this heterogeneity may actually be its greatest strength.

Resilience as theory and prescription

The varying conceptualisations of resilience have been summarised and their origins explained elsewhere and so will not be repeated at length here (see for example, Cretney, 2014; Davoudi, 2012; Folke et al., 2010; Pike et al., 2010). For the purposes of my arguments, however, these interpretations can be broadly summarised and simplified into three types which relate to the various definitions of the term used by those involved in adaptation-related policymaking (Davoudi et al., 2013; Manyena, 2006):

Engineering resilience is a measure of the speed of a system to return to its equilibrium state. The faster the system bounces back, the more resilient it is (Holling, 1973). When applied to contemporary social-ecological interactions, the assumption associated with this interpretation appears in practice to be that human society and its individual communities are discrete entities whose stability and structure we must endeavour to protect and maintain in the face of external pressures.

Ecological resilience, on the other hand, relates to both the magnitude of disturbance that can be absorbed within a system's zone of stability and the speed of recovery or adjustment to *one of a number of alternative equilibrium states*. Ecological resilience is measured by the ability to both persist and adapt. Equilibrium states are focused on system dynamics rather than structural stability. This characterisation, developed by Holling (1996) amongst others, was the precursor to the theory of social-ecological systems as compromising a *panarchy*[4] of co-dependent organisms and ecosystems that are prone to shifting between varying equilibrium states depending on both the endogenous and exogenous pressures placed upon them (Kirchhoff et al., 2010; Gunderson and Holling, 2002).

Evolutionary resilience borrows largely from the idea of dynamical social-ecological adaptive panarchies (Davoudi, 2012), but is often promoted from a social-scientific point of view that doesn't necessarily subscribe to panarchy as an ontological premise (Welsh, 2014). Evolutionary resilience largely eschews the idea of equilibrium states. It is premised on the idea that change is a fundamental component of SESs irrespective of any given external disturbance. Evolutionary resilience is therefore associated with the ability of communities to demonstrate agency, to adapt, respond and transform in response to stresses and strains and the inevitable changeability of SESs. Evolutionary resilience is based on the combined attributes of stability and flexibility in the face of intractable uncertainty and inevitable change. As Davoudi (2012, p. 302) explains, evolutionary resilience is based on a view of the world as chaotic, uncertain and unpredictable:

> Faced with adversities, we hardly ever return to where we were … regime shifts are not necessarily the outcome of an external disturbance and its linear and proportional cause and effects … change can happen because of internal stresses with no proportional or linear relationship between cause and effect.

Much of the criticism of resilience as a normative prescription has been levelled at social-ecological views that promulgate an equilibrium-based perspective on the society-environment relationship. On a purely practical level some argue that the current emphasis of climate adaptation policies and disaster risk management on 'bouncing back' (engineering resilience) or 'bouncing forth' (ecological resilience) are problematic because they assume an ability to optimise governance interventions to maintain equilibrium states in a way that fails to adequately account for the need for transformative change (Welsh, 2014; Funfgeld and McEvoy, 2012). Davoudi et al. (2013) describe how climate change adaptation plans in England have demonstrated an interpretation of resilience as, at best, ecological and at worst, engineering. These equilibrium-based interpretations frame the approach of many advanced economies to climate adaptation, in ways that emphasise a need to reduce climate risk by eliminating exposure and vulnerability to future sudden, large and turbulent events.

This equilibrium-based view also relates to dominant formulations of climate change adaptation problems as a practice of '*predict-then-act*' climate risk management, still often promoted by climate change scholars (see Church et al., 2016). Under this view climate risks can be empirically and reliably assessed and prioritised and a well-adapted community is deemed to be one that can perpetuate indefinitely its existing socio-economic structures and institutions. This reductionist conceptualisation appears to deny the partiality and contingency of climate risks described in previous chapters, as well as the fundamentally changeable nature of SESs (Folke, 2010; Scheffer, 2009). Furthermore, as argued by Pike et al. (2010), equilibrium-based approaches to climate risk management do not adequately account for the geographical and political diversity and unevenness of communities' resilience, nor adequately address questions concerning the resilience 'of what and for whom'.

Others have criticised social-ecological views of resilience on the basis that these concepts make unwise presuppositions about the panarchical, nested, ecosystems-based model of the natural environment as having a basis in objective reality, rather than being merely a prevailing conceptual model of how organisms and ecosystems interact (Welsh, 2014; Kirchhoff et al., 2010). Such presuppositions appear to ignore ongoing debates about the objective veracity of this social-ecological model compared to competing ideas that argue for ecological individualism, whereby organisms and species respond individualistically to ecosystem perturbations, and neither reach for nor move toward equilibrium states (Kirchhoff et al., 2010). The ecological view of resilience, it is argued, suggests that human society is bound by the same dynamical and behavioural principles as ecosystems and other species and not capable of combining in a wide variety of ways that may be more or less in tune with nature (Hornborg, 2013; Kirchhoff et al., 2010). By implication, this model fails to allow for the agency of individuals and communities in effectively responding to change by shaping their adaptation through transformation (Cretney, 2014).

The main problem with equilibrium-based views of resilience (as well as the risk assessments conducted on the back of such assumptions), I argue, is that because they often perceive SESs as structurally static (Gunderson and Holling, 2002: p. 26), and adaptation as the practice of preservation and/or conservation of society, economy and ecology, they undervalue flexibility as an important attribute of community, institutional and ecological resilience, relative to their ongoing stability (Pike et al., 2010). By contrast, evolutionary resilience can be maintained and enhanced by two alternating and potentially conflicting approaches that appear to align well with the contrasting goals of societal stability and flexibility.

The first approach, *adaptation*, aligns with efforts to conserve, to maintain existing infrastructure and institutions and the (political and practical) need for short-term responses to extant and expected hazards. However, adaptation may limit the possibility of, and the flexibility required for, transformative change as a result of the path-dependency of particular engineering and infrastructure approaches (Pike et al., 2010) (see my typology above, and adaptation's wicked characteristics in Chapter 4); and as a result of incremental policymaking and the political desire to build upon existing policies (Levin et al., 2010; EC 2009; COAG 2007). The second approach to maintaining evolutionary resilience, *adaptability*, or adaptive capacity, recognises the potential dangers associated with path-dependent incrementalism, and therefore the need for mechanisms to allow policymakers to break free from policy constraints and norms where necessary, in order to evolve with the chaotic and ever-changing nature of SESs. Adaptability is concerned with the development of community capacity and institutional flexibility to adjust, and where necessary transform,[5] in response to inevitable and unpredictable change, while maintaining some fundamental function of communities (Pike et al., 2010).

Some have criticised resilience as a normative concept more broadly, however, on the basis that it plays to a neo-liberal ideology, by implicitly arguing for the rolling back of the state and the primacy of individual and corporate interests (Joseph, 2013; MacKinnon and Derickson, 2012; Peck and Tickell, 2002). In this way, rather than assuming an ability to predict and optimise adaptation responses, resilience would be used as a means to advocate for 'bottom-up' approaches that would reduce the responsibilities of government for climate risk management. Indeed, I agree that, as a normative prescription, we must be careful about how resilience is (mis)used by government. As I argue below, this is exactly why the concept holds significant strength as the basis of an epistemic framework since it allows for a degree of normative transparency and therefore critical analysis of the evidence and rhetoric produced by government.

From an alternative perspective, however, the implied – and sometimes explicit (Joseph, 2013) – stance underpinning much of the criticism of resilience as a neo-liberal prescription seems to be that government can and should seek to minimise any and all climate risks to its citizens, and in this regard, some caution is advised. There is a pragmatism needed here, I believe, concerning policy players' *known-* and *unknown-unknowns* and both the political and practical feasibility of pre-emptive policy measures to address them. The nature of the scientific uncertainties associated with predict-then-act approaches to climate adaptation (see Chapter 4) should be recognised by practitioners when considering the concept of resilience and developing understandings of climate extremes and change (Chandler, 2014).

In practice it is not possible to adequately *predict* (i.e., in a credible and salient way) the extent and dynamics of future climate events or to eliminate climate risk for all communities through government intervention. Attempts to eliminate risk often entail large public expenditure under considerable uncertainty (Frigg et al., 2015). Examples where such costs are deemed justified – such as, for example, flood defences for the city of London (EA, 2012b) (see Box 4.3, Chapter 4), or the Wivenhoe dam in southeast Queensland (QFCI, 2011) – may inevitably and necessarily be the exception rather than the rule, highlighting as they do the contingencies and political inequalities associated with climate adaptation policies. They are also usually justified on the basis of existing climate vulnerability rather than in anticipation of future climate change, irrespective of prevailing socio-political ideology (Tangney and Howes, 2016); and, may be found wanting in any case (see Box 4.4, Chapter 4). Therefore, although we should be concerned by potential uses of resilience as a means to promulgate neo-liberal values that might abdicate government responsibility for climate adaptation, there is also a pragmatic and increasing necessity for climate adaptation that looks beyond state-sponsored infrastructure provisions and understands the need for community agency in preparing for and responding to change. Indeed, some argue that, alongside the co-opting of resilience by neo-liberal forces, the concept has also been taken up by grassroots community organisations as a means to encourage counter-cultural activism and community adaptive capacity (Cretney, 2014; Welsh, 2014) and shows some potential in this regard (Chandler, 2014; Hornborg, 2013).

For the purposes of adaptation policy, and in light of the various criticisms detailed above, it would seem that an *evolutionary* view of resilience appears most progressive as a prescription for the development of diverse and context-specific solutions to climate change, while simultaneously providing strategic coherence for policymakers (Davoudi et al., 2013). By allowing for ideas of ecological individualism and eschewing the idea of equilibrium states, the evolutionary view can account for any potential lack of coherence or co-dependency between natural and social systems (Hornborg, 2013) and accounts for the agency of individuals and communities to effect transformation in the face of extreme climate (Cretney, 2014).

Resilience as reflexive epistemic framework

Irrespective of which theoretical or normative formulation is preferred, when used as the foundation of an epistemic framework, I argue, resilience can account for a plurality of political and social-ecological perspectives and cuts to the heart of the issue of climate adaptation. Like risk, resilience is a relatively straight-forward concept at face value, while varying conceptions of resilience described here directly address many of the underlying norms and values of policy players, relating to what is valued and by whom, when formulating and managing adaptation problems. Under a resilience-based evidence framework, policymakers' ideals may still be partially implicit but I believe they can be considerably more transparent than through a risk-based framework. As shown in Chapter 7, risk-based decision-making often leaves political and normative ideals concealed behind a linear-technocratic conceit, and risks are so often conceptualised and circumscribed in reductionist scientific terms of likelihood, consequence and/or expert judgement that do not require the expression of underlying normative positions or the exploration of problem contingency.

Resilience, on the other hand, through its conceptualisation of social-ecological systems reveals a measure of its own contingency beyond linear-technocratic ideals. In order to ensure adequate legitimacy, credibility and salience, the resilience-based approach inevitably forces a knowledge system to address questions concerning the resilience of what and for whom, since no empirical, scientistic formula or expert 'confidence level' can adequately replace or distract from such important value positions. This suggests that a resilience-based frame can make the normative components of evidence development considerably more transparent. Resilience-based approaches may, therefore, provide a suitable framework through which evidence development can avoid the types of politicisation and scientisation facilitated by linear-technocratic methods.

Some have criticised resilience on the basis that its heterogeneity amounts to a meaningless concept (Joseph, 2013). In the context of linear-technocratic governance models, however, I argue that its plurality of potentially valid meanings can be used for the benefit of evidence-based policymaking. The value of resilience as the foundation of policy evidence development is the same characteristic that makes it difficult to characterise as social-ecological theory; it can be interpreted in a variety of ways to account for contrasting values and political priorities. For instance, Davidson et al. (2016) argue that a key dependent variable that differentiates between alternative conceptualisations of resilience, is the level of system disturbance they are geared towards. Whereas engineering resilience may be able to account for disturbances to stable socio-economic systems and climate regimes, the evolutionary approach speaks to a need to transform in the face of game-changing climate extremes. In this way, resilience-based policy evidence could potentially inform about government's relative attitudes to the prospect of future climate change, relative to existing climate variability. From a policymaking and political perspective, however, alternative concepts of resilience equate not just to the relative exposure to, or appetite for, climate hazards and system disturbance, but also to the prevailing appetites for socio-political and economic conservatism of the political elite (MacKinnon and Derickson, 2012).

Resilience, as a heterogeneous concept that accounts for varying political interpretations of the permanence and importance of socio-economic and governance structures, therefore, may allow for the derivation of adaptation policies in terms that can satisfy varying levels of politico-economic conservatism and legitimacy for climate change as an extant policy priority. At the same time, the use of this concept avoids problematic linear-technocratic notions of being able to definitively apprehend adaptation problems through objective policy evidence because it does not prescribe positivistic risk assessment or definitive expert judgement which has been so problematic during policy evidence development for climate change to date.

The linear-technocratic schema associated with risk-based adaptation has largely downplayed and even denied the inevitable normative components of risk assessment and has facilitated processes of politicisation and scientisation as a result. Moreover, risk-based approaches align with problematic notions that climate projections can make decision-relevant predictions about future climate in order to optimise adaptation responses (Frigg et al., 2013b, 2015). A credible and salient resilience-based approach may still incorporate empirical assessment of climate risk, where appropriate; for instance, as one component of a normatively transparent assessment, but it would do so in a way that recognises intractable uncertainties and the inherent contingency of risk determinations (Chandler, 2014). Likewise, resilience does not necessarily make the use of climate change projections and scenarios redundant in the way that a simple climate vulnerability assessment may be seen to. Like vulnerability, however, resilience speaks to the contingent and partial nature of adaptation problems. As such, a resilience-based framing of adaptation problems may provide an optimal combination of risk and vulnerability (or 'top-down' versus 'bottom-up') approaches to evidence development, by embracing the contingency of policy objectives and limiting processes of politicisation, while ensuring that the best available climate science can still be used to understand the competing demands for the stability and flexibility of social-ecological systems.

Resilience as a governance concept is very much under development and therefore still hotly contested. There are multiple ways to use it as a prescription for the management of society, economy, climate and the natural environment. And so there should be. We may never have a definitive prescription of resilience, or even, adequate understanding of social-ecological dynamics that would allow a precise definition. What is needed is a greater understanding of how it may be used as a frame of governance, to assist both evidence and policy development. My hope, therefore, is that resilience will at least spark further transdisciplinary debate concerning *how* we seek to know about climate adaptation and the management of climate change.

Conclusion

In this chapter I have highlighted the ways in which uncertainty manifests itself when policy players seek to address adaptation policy problems within a particular context, and from a particular government position. These uncertainties suggest that the evidence needs and contingencies of policy players may vary considerably depending on any particular problem, context or governance level. The typology used here, however, also points to the ongoing challenges for expert and non-expert policy players when preparing salient evidence for policymaking. Although addressing the uncertainties associated with government goals and objectives can be unpalatable or politically unacceptable, failing to do so may result in evidence-based policies and plans falling into a linear-technocratic trap that would foment processes of politicisation and scientisation. Risk-based approaches to climate adaptation in particular, that assume an ability to empirically, adequately and impartially apprehend future climate risks, or that are justified through recourse to independent expert authority, are particularly susceptible to these processes.

A resilience-based approach that necessitates a degree of normative characterisation of adaptation problems even when not stated explicitly; that can account for varying levels of political-economic conservatism on the part of the political executive; and that can incorporate but transcend assessments of climate change risk and vulnerability, may be a viable and worthy alternative. In Chapter 9, I discuss further how knowledge systems can incorporate evidence in a way that might overcome the tensions between politics and expertise. Although there may be no easy answers in this regard, an important first step, I argue, is to understand the limitations to existing expertise on climate change and to find a way for evidence to be developed on the basis of explicit values and priorities.

Notes

1   Kirchoff et al. (2013) cite Nobel Laureate Daniel Kahneman's (2011) seminal work relating to the distinction between 'coherence' and 'correspondence' forms of cognition, summarised in the book *Thinking, Fast and Slow*.

2   I was lead assessor of the Environment Agency's climate change risks, opportunities and responses for this report, as well as principal author of the Agency's final report to government.

3   The shifting character of uncertainty can be seen, for instance, by comparing the Environment Agency's 2010 risk assessment used here, the UKCCRA (2012) discussed in Chapter 7, and the recently published UKCCRA (2017) (ASC, 2016).

4   A term coined by Holling referring to the adaptive nature of social-ecological systems, which alludes to the Greek god Pan and to the hierarchical and nested interconnections between the adaptive cycles of ecosystems and their constituent species (Gunderson and Holling, 2002).

5   Walker et al. 2004 argue there is a significant distinction between the tasks of adaptability and transformability, but also note that the social capital required for both are very similar.

Bibliography

Adaptation Sub-Committee [ASC] 2016, *UK Climate Change Risk Assessment 2017 Synthesis Report: Priorities for the Next Five Years*, Adaptation Sub-Committee of the Committee on Climate Change, London.

Adger, WN, Brown, K, Nelson, DR, Berkes, F, Eakin, H, Folke, C, Galvin, K, Gunderson, L, Goulden, M, O' Brien, K, Ruitenbeek, J and Tompkins, EL 2011, 'Resilience implications of policy responses to climate change', *WIRES Climate Change*, vol. 2, no. 5, pp. 757–766.

Australian Government 2007, *Climate Change Adaptation Actions for Local Government*, Australian Greenhouse Office, Department of the Environment and Water Resources, Canberra, viewed 25 August 2013, www.climatechange.gov.au/en/what-you-can-do/community/~/media/publications/local-govt/localadaption_localgovernment.pdf.

Carpenter, SR, Westley, F and Turner, MG 2005, 'Surrogates for resilience of social-ecological systems', *Ecosystems*, vol. 8, pp. 941–944.

Chandler, D 2014, 'Beyond neoliberalism: Resilience, the new art of governing complexity', *Resilience*, vol. 2, no. 1, pp. 47–63.

Church, J, Hobday, A, Lenton, A and Rintoul, S 2016, 'Six burning questions for climate science to answer post-Paris', *The Conservation*, 29 February 2016, viewed 19 August 2016, https://theconversation.com/six-burning-questions-for-climate-science-to-answer-post-paris-55390/.

Council of Australian Governments [COAG] 2007, *National Climate Change Adaptation Framework, Council of Australian Governments*, Commonwealth of Australia, viewed 9 September 2012, www.climatechange.gov.au/government/initiatives/national-climate-change-adaptation-framework.aspx/.

Council of Australian Governments [COAG], 2011. *National Strategy for Disaster Resilience: Building Our Nation's Resilience to Disasters*, National Emergency Management Committee (NEMC) Working Group tasked by the COAG, Commonwealth of Australia, viewed 11 April 2013, www.coag.gov.au/.

Cretney, R 2014, 'Resilience for whom? Emerging critical geographies of socio-ecological resilience', *Geography Compass*, vol. 8, no. 9, pp. 627–640.

Davidson, JL, Jacobson, C, Lyth, A, Dederkorkut-Howes, A, Baldwin, CL, Ellison, JC, Holbrook, NJ, Howes, MJ, Serrao-Neumann, S, Singh-Peterson, L and Smith TF 2016, 'Interrogating resilience: Toward a typology to improve its operationalization', *Ecology and Society*, vol. 21, no. 2, pp. 27.

Davoudi, S 2012, 'Resilience: A bridging concept or a dead end?', *Planning Theory & Practice*, vol. 13, no. 2, pp. 299–333.

Davoudi, S, Brooks, E and Mehmood, A 2013, 'Evolutionary resilience and strategies for climate adaptation', *Planning, Practice & Research*, vol. 28, no. 3, pp. 307–322.

Department of Environment, Food and Rural Affairs [DEFRA] 2009, *Adapting to Climate Change: Helping Key Sectors to Adapt to Climate Change*, DEFRA, London, viewed 11 March 2013, http://archive.defra.gov.uk/environment/climate/documents/interim2/report-guidance.pdf.

Dessai, S and Hulme, M 2004, 'Does climate adaptation policy need probabilities?', *Climate Policy*, vol. 4, no. 2, pp. 107–128.

Duit, A, Galaz, V, Eckerberg, K and Ebbeson, J 2010, 'Governance, complexity, and resilience', *Global Environmental Change*, vol. 20, no. 3, pp. 363–368.

Emergency Management Australia [EMA] 2004, *Emergency Management in Australia: Concepts and Principles*, Attorney General's Department, Commonwealth of Australia, Canberra, viewed online 17 March 2015, www.em.gov.au/Documents/Manual01-Emergency ManagementinAustralia-ConceptsandPrinciples.pdf.

Environment Agency of England & Wales [EA] 2010a, *Managing the Environment in a Changing Climate*, EA, Bristol, UK, viewed 17 March 2015, www.environment-agency. gov.uk/research/library/publications/130528.aspx/.

European Commission 2009, *White Paper: Adapting to Climate Change: Towards a European Framework for Action*, EC, Brussels, COM(2009) 147 final, viewed 17 March 2015, http://eur-lex.europa.eu/LexUriServ/LexUriServ.do?uri=COM:2009:0147:FIN:EN: PDF.

Folke, C, Carpenter, SR, Walker, B, Scheffer, M, Chapin, T and Rockstrom, J 2010, 'Resilience thinking: Integrating resilience, adaptability and transformability', *Ecology and Society*, vol. 15, no. 4, pp. 20.

Folke, C, Carpenter, SR, Walker, B, Scheffer, M, Chapin, T and Rockstrom, J 2010, 'Resilience thinking: Integrating resilience, adaptability and transformability', *Ecology and Society*, vol. 15, no. 4, pp. 20.

Frigg, R, Smith, LA and Stainforth, DA 2013b, 'The myopia of imperfect climate models: The case of UKCP09', *Philosophy of Science*, vol. 80, no. 5, pp. 886–897.

Frigg, R, Smith LA and Stainforth DA 2015. 'An assessment of the foundational assumptions in high-resolution climate projections: The case of the UKCP09', *Synthese*, 192, no. 12, pp. 3979–4008.

Funfgeld, H and McEvoy, D 2012, 'Resilience as a useful concept for climate change adaptation?', *Planning Theory & Practice*, vol. 13, no. 2, pp. 324–333.

Grundmann, R 2006, 'Ozone and climate: Scientific consensus and leadership', *Science, Technology & Human Values*, vol. 31, no. 1, pp. 73–101.

Gunderson, LH and Holling, CS 2002, *Panarchy: Understanding Transitions in Human and Natural Systems*, Island Press, Washington DC.

Heazle, M, Tangney, P, Burton, P, Howes, M, Grant-Smith, D, Reis, K, Bosomworth, K 2013, 'Mainstreaming climate change adaptation: An incremental approach to disaster risk management in Australia', *Environmental Science & Policy*, vol. 33, pp. 162–170.

Holling, CS 1973, 'Resilience and stability of ecological systems', *Annual Review of Ecology and Systematics*, vol. 4, pp. 1–23.

Holling, CS 1996, 'Engineering resilience versus ecological resilience', in Schulze, P (ed.), *Engineering within Ecological Constraints*, National Academy Press, Washington DC.

Hornborg, A 2013, 'Revelations of resilience: From the ideological disarmament of disaster to the revolutionary implications of (p)anarchy', *Resilience*, vol. 1, no. 2, pp. 116–129.

Howes, M, Tangney, P, Reis, K, Grant-Smith, D, Heazle, M, Bosomworth, K and Burton, P 2015, 'Towards networked governance: Improving interagency communication and collaboration for disaster risk management and climate change adaptation in Australia', *Journal of Environmental Planning and Management*, vol. 58, no. 5, pp. 757–776.

Joseph, J 2013, 'Resilience as embedded neoliberalism: A governmentality approach', *Resilience*, vol. 1, no. 1, pp. 38 – 52.

Kahneman, D 2011, *Thinking, Fast and Slow*, Farrar, Straus and Giroux, New York.

Kirchhoff, T, Brand, FS, Hoheisel, D and Grimm, V 2010, 'The one-sidedness and cultural bias of the resilience approach', *Gaia*, vol. 19, no. 1, pp. 25–32.

Kirchoff, CJ, Lemos, MC and Dessai, S 2013, 'Actionable knowledge for environmental decision-making: Broadening the usability of climate science', *Annu. Rev. Environ. Resour.*, vol. 38, pp. 393–414.

Kuhlicke, C, Steinfuhrer, A, Begg, C, Bianchizza, C, Brundl, M, Buchecker, M, De Marchi, B, Di Masso Tarditti, M, Hoppner, C, Komac, B, Lemkow, L, Luther, J, McCarthy, S, Pellizoni, L, Renn, O, Scolobig, A, Supramaniam, M, Tapsell, S, Wachinger, G, Walker, G, Whittle, R, Zorn, M and Faulkner, H 2011, 'Perspectives on social capacity building for natural hazards: Outlining an emerging field of research and practice in Europe', *Environmental Science & Policy*, vol. 14, pp. 804–814.

Levin, K, Cashore, B, Bernstein, S and Auld, G 2010, 'Playing it forward: Path dependency, progressive incrementalism, and the 'super wicked' problem of global climate change', International Studies Association Convention, 28 February – 3 March 2010, Chicago, IL, viewed 17 March 2015, http://citation.allacademic.com/meta/p_mla_apa_research_citation/1/7/9/7/0/pages179707/p179707-1.php/.

MacKinnon, D and Derickson, KD 2012, 'From resilience to resourcefulness: A critique of resilience policy and activism', *Progress in Human Geography*, vol. 37, no. 2, pp. 253–270.

Maguire, B and Hagan, P 2007, 'Disasters and communities: Understanding social resilience', *The Australian Journal of Emergency Management*, vol. 22, no. 2, pp. 16–20.

Manyena, SB 2006, 'The concept of resilience revisited', *Disasters*, vol. 30, no. 4, pp. 433–450.

Meyer, R 2011, 'The public values failures of climate science in the US', *Minerva*, vol. 49, no. 1, pp. 47–70.

Moss, RH, Meehl, GA, Lemos, MC, Smith, JB, Arnold, JR, Arnott, JC, Behar, D, Brasseur, GP, Broomell, SB, Busalacchi, AJ, Dessai, S, Ebi, KL, Edmonds, JA, Furlow, J, Goddard, L, Hartmann, HC, Hurrell, JW, Katzenberger, JW, Liverman, DM, Mote, PW, Moser, SC, Kumar, A, Pulwarty, RS, Seyller, EA, Turner II, BL, Washington, WM and Wilbanks, TJ 2013, 'Hell and high water: Practice-relevant adaptation science', *Science*, vol. 342, pp. no. 6159, pp. 696–698.

National Emergency Management Committee [NEMC] 2010, *National Emergency Risk Assessment Guidelines*, Attorney General's Department, Commonwealth of Australia, Canberra.

Peck, J and Tickell, A 2002, 'Neoliberalizing space', *Antipode*, vol. 34, no. 3, pp. 380–404.

Pike, A, Dawley, S and Tomaney, J 2010, 'Resilience, adaptation and adaptability', *Cambridge Journal of Regions, Economy and Society*, vol. 3, no. 1, pp. 59–70.

Pitman, A 2016, 'CSIRO boss' failed logic over climate science could waste billions in taxes', *The Conversation*, 5 February 2016, viewed online 26 December 2016, https://theconversation.com/csiro-bosss-failed-logic-over-climate-science-could-waste-billions-in-taxes-54249.

Preston, BL, Rickards L, Dessai S and Meyer, R 2013, 'Water, Seas, and Wine: Science for Successful Climate Adaptation', in Moser, SC and Boykoff, MT (eds), 2013, *Successful Adaptation to Climate Change: Linking Science and Policy in a Rapidly Changing World*, Routledge, Abingdon.

Prosser, B and Peters, C 2010, 'Directions in disaster resilience policy', *The Australian Journal of Emergency Management*, vol. 25, no. 3, pp. 8–11.

Queensland Floods Commission of Inquiry [QFCI] 2011.*Queensland Floods Commission of Inquiry: Interim Report*, QFCI, Brisbane, viewed 3 April 2013, www.floodcommission. qld.gov.au/publications/interimreport/.

Queensland Government 2012b, *Queensland Local Disaster Management Guidelines*, Emergency Management Queensland, viewed 25 August 2014, www.disaster.qld.gov. au/Disaster-Resources/Documents/Queensland%20Local%20Disaster%20 Management%20Guidelines.pdf.

Risk Frontiers 2012, *Historical Analysis of Natural Hazard Building Losses and Fatalities for Queensland 1900–2011*, Queensland Department of Community Safety, viewed 21 August 2014, http://disaster.qld.gov.au/Disaster-Resources/Documents/Historical%20 analysis%20of%20natural%20hazard%20building%20losses%20and%20fatalities%20 for%20Queensland%201900-2011.pdf.

Scheffer, M 2009, *Critical Transitions in Nature and Society*, Princeton University Press, Princeton, NJ.

Simmie, J and Martin, R 2010, 'The economic resilience of regions: Towards an evolutionary approach, *Cambridge Journal of Regions, Economy and Society*, vol. 3, no. 1, pp. 27–43.

Tangney, P and M. Howes 2016. 'The politics of evidence-based policy: A comparative analysis of climate adaptation in Australia and the UK', *Environment and Planning C: Government and Policy*, vol. 34, no. 6, pp. 1115–1134.

Walker, B, Holling, CS, Carpenter, SR and Kinzig, A 2004, 'Resilience, adaptability and transformability in social-ecological systems', *Ecology and Society*, vol. 9, no. 2, p. 5.

Welsh, M 2014, 'Resilience and responsibility: Governing uncertainty in a complex world', *The Geographical Journal*, vol. 180, no. 1, pp. 15–26.

Willows, R and Connell, R (eds) 2003, *Climate Adaptation: Risk, Uncertainty and Decision-Making*, UK Climate Impacts Programme, Oxford.

Yohe, G and Tol, RSJ 2002, 'Indicators for social and economic coping capacity: Moving toward a working definition of adaptive capacity', *Global Environmental Change*, vol. 12, pp. 25–40.

9 Reconciling tensions between experts, evidence and politics

Science is a first rate piece of furniture for a man's upper chamber, if he has common sense on the ground floor.

Oliver Wendell Holmes, Sr

In this book I have argued that the value of evidence for climate adaptation policymaking is determined not just (or even mostly) by its technical adequacy. For better or worse, evidence use is principally determined by how the values and political priorities attributed to or inscribed within this information align with the norms and priorities of the policymaking community. Evidence legitimacy, therefore, holds primacy over its technical credibility and instrumental salience for adaptation policymaking.

Unfortunately, the science concerning climate change adaptation has encountered serious impediments to its use for decision-making in the ways traditionally expected by policymakers and the public. Scientists cannot provide decision-relevant forecasts of future climate to the satisfaction of policymakers, and policy players increasingly perceive how climate change projections are often too uncertain, complex, inappropriately designed or contentious to be able to linearly or prescriptively inform policy decisions, particularly when competing with neo-liberal economic-rationalist norms relating to the establishment of a business case for adaptation action. So much so, in fact, that many policy players see climate change projections as a secondary concern in the development of robust evidence for climate adaptation. Instead, information relating to the vulnerability of relevant jurisdictions and communities, or of other facets of socio-economic or ecological systems has often been a priority focus. Moreover, climate change impacts and associated adaptation problems resist definitive characterisation and depend upon the perspective from which they are assessed. Even so, evidence concerning the impacts from and vulnerability to climate change is still often being shoe-horned into risk-based methods of evidence interpretation and assessment for policymaking alongside climate projections, in ways that assume that this information can be provided in a positivist and reductive way.

Risk-based methods generally ascribe to the notion that climate risks can be impartially and empirically assessed, primarily with the use of scientific and other technical expertise. Yet, such knowledge, particularly at strategic governance levels, is necessarily uncertain and contingent and cannot be provided impartially to satisfy all (or even most) relevant stakeholders. Climate change impacts (and therefore assessments of risk) depend on context and location. Many policy players with first-hand experience of the practical impediments encountered when using climate science and derived adaptation science for policymaking, therefore, increasingly advocate a pragmatic, decision-based approach. This would combine decision-makers' transparent values and priorities alongside both expert and non-expert understandings of vulnerability and resilience, to derive legitimate technical interpretations of evidence for climate adaptation. This approach would counteract long-held expectations that policy can and should, first and foremost, be informed by independent, objective experts and their privileged knowledge of climate change.

Positivist linear-technocratic views aligning with ambitions to be able to predict future climate change and, on that basis, to definitively apprehend its potential impacts and risks, are fundamentally wrong-headed. Although climate change risks should be taken very seriously, climate models will likely never be able to reliably predict the future, and impact assessments are necessarily dependent on a range of assumptions and contingencies that make their resolution of system complexities necessarily partial and uncertain. Moreover, both experts and policymakers are often reluctant to be explicit about their values, ideals and objectives in the context of evidence development for adaptation policy. For fear of not being objective, therefore, they often leave important normative judgements implicit and even covertly expressed through the development, dissemination and use of expert knowledge. As proposed in Chapter 4, a politicisation-by-process can therefore occur through the development of trans-science; through post-normal processes of co-design of scientific outputs; and further, through the co-production of subsequent policy evidence. And still, governments cling to linear-technocratic ideals. Society expects expert evidence to be objective, transparent and true, yet meaningful information about climate adaptation policy problems – such as understanding the impacts and risks from climate change – necessitates significant value judgements during its derivation, thus ensuring that this knowledge will nearly always be contingent.

As discussed in preceding chapters, as scientific research approaches the science-policy interface, it requires increasing levels of subjective and normative interpretation to ensure it is salient and politically acceptable for a chosen set of policy players. The interpretive input required in the development of policy evidence also means that, although its credibility, legitimacy and salience may be sufficient for those who produced it and for whom it is directly produced, this information can often be perceived as lacking usability, as being technically deficient or irrelevant, or even politically derived by other groups of policy players instrumental in implementing effective policies. In particular, those with conflicting priorities, those situated at different levels of government, or those who are concerned with differing governance scales may reject the available policy evidence due to its necessarily contingent nature.

More sinister than the inevitable socio-cultural influences upon expertise, however, are processes of deliberate politicisation that are facilitated by society's expectations for the privileged objectivity of experts when informing policymaking. This politicisation-by-agency occurs when policymakers and other players seek to suppress explicit normative debate concerning important social and environmental risks, and thus to rationalise government policies through politicised knowledge. Both forms of politicisation discussed here occur as a result of the tensions arising from the inevitable overlap between expert and political authority in public decision-making. These tensions appear at their zenith for hyper-politicised 'wicked' problems such as climate change.

Some key questions

How do contextual and political forces influence the role of experts in climate adaptation?

In Chapter 4 I explained the challenges of evidence provision for climate adaptation, and climate change in particular, which mean that processes of knowledge co-design and co-production between certified and non-certified experts and non-experts are both necessary and inevitable in order to make climate science legitimate and salient, and to allow for the development of useful policy evidence. Calls for the democratisation of expertise relate to expectations of access for a greater number of participants and the public to expert evidence, and even a revised definition of expertise itself, whereby local and contextual knowledge from so-called 'lay experts' have become a valid contribution alongside science and economics. Legitimate expertise, under this view, is an attribute that has been extended to a much broader cohort of policy stakeholders, and contextual and political forces are necessary and inevitable constituents of expertise for climate adaptation policymaking.

Evidence co-design and co-production processes appeal to aspirations for an enhanced legitimacy for policy evidence, even though they simultaneously contradict society's linear-technocratic expectations and are problematic when adhering to the guidelines of liberal-democratic governance that prescribe positivist notions of robust, credible and impartial policy evidence. However, as shown here, perspectives on climate hazards, impacts, risks and responses will vary depending on the policy players involved and are subject to a wide variety of valid expert interpretations. Technical experts called upon to provide synoptic knowledge or judgement subsequently struggle to fulfil their traditional roles as the purveyors of objective or immutable truths for policymaking, and indeed may also fail to fulfil a more progressive ideal for the provision of impartial judgement on issues that transcend anyone's precise expertise, particularly for those issues like climate change that are highly politically contentious.

The comparative political analysis in Chapter 3, alongside the collected views of adaptation policy players summarised in Chapter 6, suggest that contextual and political forces have an important and inevitable influence on experts and evidence due to the 'wicked' characteristics of adaptation policy problems. Under a linear-technocratic schema, experts have the opportunity to extend their authority beyond that which they can legitimately claim privileged knowledge about. Alternatively, the façade of objective expertise may facilitate the politicisation of policy evidence by non-experts participating in co-production processes.

These findings suggest that experts should continually reframe and restate the precise limits of their expertise for policymaking while being explicit about the contingencies involved. The development of evidence requires experts to provide commentary and guidance on a range of issues related to, but not confined within, their particular area of privileged knowledge. Engaging in the policy evidence development process requires experts to engage in both technical and a socio-cultural reasoning, the latter indicating that experts must be politically in tune with the policymaking community and realising that their privileged knowledge of adaptation issues has important limits. Even within their sphere of direct expertise, experts' contributions may lack legitimacy and salience when they fail to align with the evolving politics of what is considered useful, usable and politically acceptable. Although not traditionally a core strength of scientific experts, I believe it is time for scientists to learn the difference between their epistemic and non-epistemic value judgements and to understand the importance of linking these various judgements to the political context of their roles as policy advisers.

However, if policymaking labours under false expectations for objective evidence and impartial expertise, why then don't we seek to remove such unrealistic expectations and to recalibrate how we perceive evidence and expertise for policy? The difficulties that arise in relation to such a shift concern, I believe, the problems of legitimacy and extension of expert authority (see Chapter 2) and the apparent political expediency of allowing bureaucratic government to answer some types of political problem without open deliberation.

Can we reconcile problems of legitimacy and extension to alleviate the tensions between expert and political authority?

The problems of legitimacy and extension described by Collins and Evans (2002) relate to finding an appropriate role for expert authority in public decision-making. Should technical decisions be informed by the widest possible range of public perspectives, or should their characterisation and resolution be informed primarily by experts? If the latter, then what counts as legitimate expertise for a problem as open, complex and contingent as climate adaptation and who should be involved in evidence development for policymaking?

Concepts of legitimacy in relation to policy knowledge have, in the past, been used in contrasting and somewhat conflicting ways. Cash et al.'s (2002) knowledge systems framework suggests that legitimacy relates principally to the knowledge production process, and not specifically to experts or to the resulting evidence itself. As discussed in Chapter 5, defining legitimacy in this way leads to a failure to distinguish between the technical credibility and political acceptability of evidence on the one hand, and between its instrumental salience and policymaking relevance on the other. Cash et al.'s (2002) perspective on evidence legitimacy, however, does align with Collins and Evans' (2002) ideas on expert legitimacy and extension, in so far as considering the legitimacy of evidence co-production processes, under a linear-technocratic schema, is dependent on the credentials of the policy players involved. Collins and Evans' (2002) ideas of expert legitimacy, likewise, seem to relate to both the credibility and political acceptability of expert policy knowledge. Nonetheless, by focusing on the knowledge production process Cash et al. appear to conflate evidence legitimacy with concurrent ideas of credible evidence production, and thereby fail to adequately account for extrinsic political influences that may affect expert knowledge during deliberative co-production processes.

In amended form, however, I argue that Cash et al.'s framework can be a valuable practitioners' tool for demonstrating the tensions between expert and political authority. It is appealingly simple in the way that it balances competing attributes of effective evidence, and in the process also touches upon the problems of legitimacy and extension of expertise prevalent during the co-production of policy evidence, without placing undue emphasis on *contributory* expertise, a concept which seems to lack a clearly definable role in the context of co-produced climate adaptation evidence development and the contingencies of adaptation problems.

However, if the evidence used for policymaking is not credible in terms of being overseen by technically competent, objective, impartial and privileged (i.e., not available to most others) expertise, then how else can we determine the credibility of policy knowledge, and what use would expert evidence be as a counterpoint to the politics of policymaking? Collin and Evans' (2002) ideas highlight this important difficulty in understanding contemporary policy problems concerning social-ecological systems and what and how we know about them, and unfortunately there are perhaps no easy answers to these questions. The one that I give, however, is that we should accept a limited scientistic framework for evidence development that nonetheless requires public oversight and co-production of the natural history of climate and how we should adapt to it.

Satisfying the need for technical credibility with the concurrent needs for political acceptability and instrumental salience, I argue, must continue to be a balancing act that strives for *robust* evidence (to be determined by individual contexts) in order to inform policy decisions. And this endeavour will likely always be susceptible to bias and politicisation by those who either fail to distinguish technical considerations from normative ones, or who seek to deliberately manipulate supposedly impartial knowledge systems for political gain. Cash et al.'s framework, with some adjustment, seems a useful tool for carefully considering these tensions, even if it fails to pin down exactly what counts as legitimate and credible *expertise* in this regard. As discussed in Chapter 5, a broader definition of legitimacy retains its relationship to the technical credibility and instrumental salience of evidence, but provides a clearer distinction across the three criteria between the technical and the political, than in its original form. Knowledge systems do not exist in a political vacuum, but are nonetheless a vital arena for managing the tensions between technical and socio-cultural knowledge.

How does the need for political legitimacy of adaptation policy influence the provision of policy evidence?

Given the need for a broad range of knowledge concerning climate risks and adaptation, and the inevitable normative/political constituents of policy evidence, in this book I have shown how, in practice, legitimacy holds primacy over credibility and salience when balancing the tensions between expert and political authority for evidence-based policymaking. Where evidence legitimacy is established, credibility is easily attained (sometimes irrespective of its technical adequacy!), and salience takes over as the principal remaining concern for policy players seeking to understand climate risks. However, where evidence legitimacy does not exist, neither credibility nor salience is sufficient to convince policy players of the suitability of this evidence for policymaking purposes.

This book demonstrates an important two-way correlation between the presence (or absence) of legitimacy for climate change policy, and a concurrent legitimacy for the available science for informing that policy. Where political legitimacy exists for addressing climate change, such as established by the Climate Change Act (2008) in the UK, the associated climate science is also promoted as a legitimate and necessary component of policymaking and is provided for policymakers with considerable attention to its usability. Yet, the case studies in this book suggest, the existence of explicit legitimacy for climate science does not guarantee any more stringent evidence-based policymaking, or for that matter, any more substantive policy action on climate *change* adaptation per se.

Even where political and evidence legitimacy exists for climate change and associated expertise at a given governance level or scale, the pluralistic nature of climate adaptation problems means that policy players at differing government levels (or who are concerned with differing governance scales) may not perceive that evidence as legitimate, irrespective of its acceptability for adaptation policymaking at the level of government at which it was commissioned and derived. The political acceptability of policy evidence is necessarily contingent upon the level of governance at which is derived.

Moreover, and somewhat ironically, legitimacy for climate science, in tandem with a linear-technocratic approach to evidence-based policymaking, may mean a greater propensity for processes of politicisation and scientisation to occur. Policy evidence for climate adaptation can be inscribed with significant normative positions and/or political decisions made during co-production processes, and then disguised behind a façade of impartial expertise during the development of *ex ante* policy assessment. In turn, this politicisation may reduce the legitimacy of particular sets of policy evidence for individual policy players or advocacy coalitions, even if climate change and the underlying climate science is considered legitimate.

Where political legitimacy for climate change does not exist, however, as in the case of Queensland and at local government level in the UK, I have shown here how legitimacy for both climate science and associated policy evidence is also lacking. Indeed climate change evidence may be entirely ignored in favour of political priorities and/or concurrent evidence sets that align more effectively with existing decision-making norms and prescriptions, even if that former evidence is outwardly accepted as technically credible.

Can government climate risk assessments adequately account for political influence and the wicked characteristics of climate adaptation problems?

Ex ante processes of evidence development, that might reconcile the credibility, legitimacy and salience of effective knowledge systems, sit at the interface between expert and political authority and combine expert knowledge, local and contextual perspectives, and the political norms and priorities of policy players involved in this process. Although, upon examination in this book, much of the analysis that occurs during *ex ante* appraisal appears to be a process of deliberation and manipulation between groups of experts and non-experts, the outputs of this process are nonetheless often portrayed as impartial independent expertise. As discussed in Chapters 2 and 4 and demonstrated in Chapters 6 and 7, climate risk assessment is therefore necessarily politicised to the extent that the framing and characterisation of climate risks and adaptation responses require subjective, normative and ultimately political input.

The case-study investigations in Chapter 7 demonstrate how climate risk assessment techniques allow important political decisions to be made during the development of evidence. Under a linear-technocratic policymaking schema, these decisions garner tacit acceptance in a way that might otherwise leave them open to protracted debate or controversy within the conventional democratic forums of governance. In such cases, adaptation policymaking, I argue, becomes *scientised* through the deliberate politicisation of policy evidence, either during the development of that evidence or subsequently during its interpretation, communication and use by policy players. In fact, as the cases examined here demonstrate, the point of demarcation between expert evidence and its presentation to policymakers has become increasingly difficult to identify, since the advent of post-normal science, and because *ex ante* assessments are considered to constitute varying levels of privileged expertise, depending on the perspectives of individual policy players or advocacy coalitions. This lack of clarity about where expertise ends and politics begins strengthens my view that politicisation has become an institutionalised component of evidence development for policy and that we need to consider experts' decision-influencing authority in this context.

As discussed in Chapter 8, climate resilience may provide a suitably reflexive epistemic framing that can soothe the tensions between expertise and politics for climate adaptation. Framing the development of policy evidence around resilience may ensure a degree of normative transparency in the political ideals and priorities of government and that can avoid linear-technocratic interpretations of risk while still using empirical scientific evidence appropriately. The concept of resilience sits in contrast to risk-based approaches that are accompanied by erroneous positivist associations concerning the objective empirical assessment of likelihood and consequence. Any explicit assessment of climate resilience necessitates consideration of important foundational questions relating to the resilience 'of what and for whom', and lays open at least some of the important contingencies associated with understanding the risks and impacts from climate change and adaptation responses. In doing so, resilience-based evidence may provide a means to standardise the input of normative perspectives into the development of policy evidence so that governments' interpretations of climate resilience tell us much about their attitudes and priorities in relation to climate change, and the socio-economic and ecological foundations of communities and their governance.

Some final reflections

The analysis provided in this book suggests that the role of experts, scientific research and policy evidence in the development of climate adaptation policy is often dependent on prevailing political attitudes toward climate science and the phenomenon of climate change. Although experts can conclusively affirm the greenhouse effect and anthropogenic influences on the global climate, they cannot definitively say how the climate may change in the future or precisely what the effect of anthropogenic influence upon it will ultimately be. The complexity and uncertainty associated with understanding the implications of climate change for policy mean that scientific expertise may always be on the back foot, and that co-production processes of evidence development involving political influence are necessary to draw meaningful conclusions from the available science.

In the UK, strict adherence to the evidence-based mandate and governments' ongoing explicit allegiance to the resolution of the climate change problem has resulted in continuing attempts to understand how climate may change in the future and a commitment by successive governments to design policy on the basis of the conclusions of climate change science, even if the available evidence fails to be used instrumentally in practice. This is because adaptation policymaking has become preoccupied by a perceived need for more and better evidence for the reduction of uncertainties. This goal has ultimately resulted in adaptation policies and plans being delegitimised by their inability to present a robust evidence-based business case in advance of uncertainty reduction. Under the linear-technocratic schema, policy evidence can also become deliberately politicised in line with government's existing and developing neo-liberal political priorities, which in some instances (e.g., in relation to the risks from sea-level rise), may conflict with the conclusions of experts. This situation raises important questions about the practical utility of existing suites of complex, contentious and uncertain climate and adaptation science that are vulnerable to such politicisation during the co-production of policy evidence.

By contrast, climate adaptation policies in Queensland have developed without the use of climate *change* science. Instead, this policy has largely focused upon more explicit political and socio-economic priorities of government alongside more tangible or deterministic forms of evidence relating to the vulnerability, exposure and resilience of communities to existing extremes and weather variability, and the elimination of extant climate risks. By prioritising political values and priorities rather than depending on the development of objective facts about the future (which don't, for the most part, actually exist) Queensland's adaptation policy appears to have largely avoided the difficulties presented by climate change science and its use by a scientised bureaucracy. And yet, Queensland's communities and emergency services are in some respects quite well adapted to climate change in terms of their ability to prepare for and respond to extreme climate conditions and potential disaster events.

While it may be intuitively appealing to seek to account for future climate in the development of public policy, the intractable uncertainties associated with understanding complex and dynamic climatic and social-ecological systems, how greenhouse gases may influence them in the future, as well as the complex and contentious nature of the available science, suggest that in practice attempts to inform adaptation policymaking in a linear-rationalist way are, in many cases, doomed to failure. The tendency toward politicisation appears to validate the explicit pursuit of political priorities over the available evidence, as in Queensland in recent years, however unpalatable those priorities may seem to those worried by climate change. I argue, therefore, that risk-based approaches to adaptation policy in particular will continue to be problematic for addressing climate change problems, due to their tendency to facilitate processes of politicisation and scientisation that significantly undermine attempts at rational evidence-based policymaking.

Adaptation policymaking needs a conceptual framework that embraces the least problematic aspects of climate science – relating to anthropogenic climate forcing to date and existing social-ecological vulnerability and resilience – alongside explicit normative prioritisation concerning political values and priorities for maintaining and enhancing that resilience. Explicit normative prioritisation can, I argue, minimise the extent of evidence politicisation to which climate change risk assessments have thus far been susceptible to.

The continued use of linear-technocratic heuristics foments unrealistic expectations about the role of experts and the value of evidence for policy. It would seem sensible, therefore, to seek to adjust the rationalist expectations of government and the public to once and for all burst the positivist myth of expertise, to account for the presence of inevitable normative decisions by experts, and to understand a broader concept of what expertise is. Such an adjustment could allow the credibility of expert authority for policymaking to be maintained on a more realistic footing and allow evidence to avoid becoming overwhelmed by covert political influence.

Yet, I wonder, would such an epistemic realignment be feasible or even desirable by policymakers or those working in bureaucratic government? The one pragmatically positive outcome arising from the politicisation of *ex ante* assessment, arguably, is that it may actually facilitate the timely and expedient delivery of policies precisely because it suppresses explicit debate under the guise of expert authority. If, as I argue here, linear-technocratic heuristics facilitate evidence politicisation and therefore the scientisation of adaptation policymaking, it could be argued that *ex ante* policy evidence already provides a pragmatic means by which to resolve the tensions between expert and political authority in a way that is reminiscent of a Rawlsian ideal for a partition between democratic deliberation and democratic decision-making (see Chapter 2). Where the evidence-based mandate is adhered to (through legislative dictate or otherwise) and evidence legitimacy exists, my research suggests that politicised *ex ante* assessments may actually provide a useful means for policymakers and other players to gain tacit acceptance for political priorities and thus avoid time-consuming and potentially intractable debates concerning society's priorities for climate resilience and adaptation.

It may be that the linear-technocratic schema has endured precisely because it provides such a pragmatic expediency in the policymaking process under an evidence-based mandate. Without the processes of politicisation and scientisation this schema facilitates, policymaking might become bogged down in endless debates concerning normative minutiae about how and what we know. Alternatively, bureaucratic policymaking would be accused of being undemocratic in its decision-making concerning a myriad of minor political/normative judgements made by unelected officials and expert advisers. Therefore, perhaps bureaucracy retains its linear-technocratic heuristics to ensure the smooth and efficient running of government. The danger, therefore, is that this pragmatic use of politicisation and scientisation allows major political decisions to slip undetected through the policymaking of bureaucratic government.

Certainly it seems that attempting to pursue a truly impartial evidence-based (or evidence-informed) approach using complex, uncertain and contentious science has proven a bridge too far for experts and their privileged policymaking authority when it comes to climate change adaptation. And yet explicit normative prioritisation during the development of policy evidence presents significant risks to both expert and political authority. Explicit priorities can threaten governments' political authority when they conflict with the expert community's strong support for addressing climate change, or as a result of the contentious and divisive nature of many adaptation problems that may cause public anger due to a perceived irrationality on the part of government decision-makers. Meanwhile, explicit prioritisation in the course of presenting expert evidence may draw undue attention to the political machinations of policy evidence development and, therefore, call into question the credibility of experts as the purveyors of objective truth or impartial judgement. It would seem that easing the tensions between expert and political authority involves risks and opportunities for both forms of authority in the policymaking process.

Bibliography

Cash, D, Clark, W, Alcock, F, Dickson, N, Eckley, N and Jager, J 2002, *Salience, Credibility, Legitimacy and Boundaries: Linking Research, Assessment and Decision Making*, John F. Kennedy School of Government, Harvard University, Faculty Research Working Papers Series, November 2002.

Collins, HM and Evans, R 2002, 'The third wave of science studies: Studies of expertise and experience', *Social Studies of Science*, vol. 32, no. 2, pp. 235–296.

Index